DISCRETE MATHEMATICS

DISCRETE MATHEMATICS

Prabhat Kr. Choudhary

ANMOL PUBLICATIONS PVT. LTD.
NEW DELHI-110 002 (INDIA)

ANMOL PUBLICATIONS PVT. LTD.

Regd. Office: 4360/4, Ansari Road, Daryaganj,
New Delhi-110002 (India)
Tel.: 23278000, 23261597, 23286875, 23255577
Fax: 91-11-23280289
Email: anmolpub@gmail.com
Visit us at: www.anmolpublications.com

Branch Office: No. 1015, Ist Main Road, BSK IIIrd Stage
IIIrd Phase, IIIrd Block, Bangalore-560 085 (India)
Tel.: 080-41723429 • Fax: 080-26723604
Email: anmolpublicationsbangalore@gmail.com

Discrete Mathematics

First Edition, 2010

ISBN 978-81-261-4392-4

PRINTED IN INDIA

Printed at Mehra Offset Press, Delhi.

Contents

Preface

The Textbook of Discrete Mathematics provides an introduction to some of the fundamental concepts in modern mathematics. Abundant examples help to explain the principles and practices of Discrete Mathematics. The book intends to cover material required by readers for whom mathematics is just a tool, as well as provide a strong foundation for mathematics majors. The vital role that Discrete Mathematics plays in computer science is strongly emphasized as well. The book is useful for students and instructors, and also software professionals. The topics covered in this book have been chosen keeping in view the knowledge required to understand the functioning of the digital computer because many of its properties can be understood and interpreted within the framework of finite mathematical systems. This book provides a compact yet rigorous treatment of the major mathematical techniques available for study in theoretical computer science. In this, each chapter has been revised thoroughly to widen the scope of topics. Several solved and unsolved questions have been added in each chapter..

The book is user-friendly having a conversational style of writing covering topics such as combinatorics, proof methods, induction, sets, theory of automata. These topics are covered with examples in each chapter which brings the clarity to most complex concepts. Specifically written for students of BE (Computer/IT), BTech (Computers/IT), MCA, MCM, MSc (Mathematics/Computer Science), the book incorporates most of the recommendations of the Curriculum Committee on Computer Science/Engineering for Discrete Mathematics.

Chapter 1

Introduction

OVERVIEW

Discrete mathematics is mathematics that deals with discrete objects. Discrete objects are those which are separated from (not connected to/distinct from) each other. Integers (aka whole numbers), rational numbers (ones that can be expressed as the quotient of two integers), automobiles, houses, people etc. are all discrete objects. On the other hand real numbers which include irrational as well as rational numbers are not discrete.

As you know between any two different real numbers there is another real number different from either of them. So they are packed without any gaps and can not be separated from their immediate neighbors. In that sense they are not discrete. In this course we will be concerned with objects such as integers, propositions, sets, relations and functions, which are all discrete. We are going to learn concepts associated with them, their properties, and relationships among them among others.

Let us first see why we want to be interested in the formal/theoretical approaches in computer science.

Some of the major reasons that we adopt formal approaches are:

1. We can handle infinity or large quantity and indefiniteness with them.
2. Results from formal approaches are reusable.

As an example, let us consider a simple problem of investment. Suppose that we invest $1,000 every year with expected return of 10% a year. How much are we going to have after 3 years, 5 years, or 10 years ? The most naive way to find that out would be the brute force calculation.

Let us see what happens to \$1,000 invested at the beginning of each year for three years. First let us consider the \$1,000 invested at the beginning of the first year. After one year it produces a return of \$100.

Thus at the beginning of the second year, \$1,100, which is equal to \$1,000 * (1 + 0.1), is invested. This \$1,100 produces \$110 at the end of the second year. Thus at the beginning of the third year we have \$1,210, which is equal to \$1,000 * (1 + 0.1)(1 + 0.1), or \$1,000 * $(1 + 0.1)^2$. After the third year this gives us \$1,000 * $(1 + 0.1)^3$.

Similarly we can see that the \$1,000 invested at the beginning of the second year produces \$1,000 * $(1 + 0.1)^2$ at the end of the third year, and the \$1,000 invested at the beginning of the third year becomes \$1,000 * (1 + 0.1).

Thus the total principal and return after three years is

\$1,000 * (1 + 0.1) + \$1,000 *$(1 + 0.1)^2$ + \$1,000 * $(1 + 0.1)^3$,

which is equal to \$3,641.

One can similarly calculate the principal and return for 5 years and for 10 years. It is, however, a long tedious calculation even with calculators. Further, what if you want to know the principal and return for some different returns than 10%, or different periods of time such as 15 years ? You would have to do all these calculations all over again.

We can avoid these tedious calculations considerably by noting the similarities in these problems and solving them in a more general way. Since all these problems ask for the result of invesing a certain amount every year for certain number of years with a certain expected annual return, we use variables, say *A*, *R* and *n*, to represent the principal newly invested every year, the return ratio, and the number of years invested, respectively.

With these symbols, the principal and return after *n* years, denoted by *S*, can be expressed as

$$S = A(1 + R) + A(1 + R)^2 + \ldots + A(1 + R)^n.$$

As well known, this S can be put into a more compact form by first computing

$$S - (1 + R)S \text{ as } S = A\,((1 + R)^{n+1} - (1 + R))/\,R.$$

Once we have it in this compact form, it is fairly easy to compute S for different values of A, R and n, though one still has to compute $(1 + R)^{n+1}$.

This simple formula represents infinitely many cases involving all different values of A, R and n. The derivation of this formula, however, involves another problem. When computing the compact form for S, $S - (1 + R)S$ was computed using $S = A(1 + R) + A(1 + R)^2 + ... + A(1 + R)^n$.

While this argument seems rigorous enough, in fact practically it is a good enough argument, when one wishes to be very rigorous, the ellipsis... in the sum for S is not considered precise. You are expected to interpret it in a certain specific way.

But it can be interpreted in a number of different ways. In fact it can mean anything. Thus if one wants to be rigorous, and absolutely sure about the correctness of the formula, one needs some other way of verifying it than using the ellipsis. Since one needs to verify it for infinitely many cases (infinitely many values of A, R and n), some kind of formal approach, abstracted away from actual numbers, is required.

Suppose now that somehow we have formally verified the formula successfully and we are absolutely sure that it is correct. It is a good idea to write a computer programme to compute that S, especially with $(1 + R)^{n+1}$ to be computed. Suppose again that we have written a programme to compute S. How can we know that the programme is correct ? As we know, there are infinitely many possible input values (that is, values of A, R and n). Obviously we can not test it for infinitely many cases. Thus we must take some formal approach.

Related to the problem of correctness of computer programs, there is the well known "Halting Problem". This problem, if put into the context of programme correctness, asks whether or not a given computer programme stops on a given input after a finite amount of time. This problem is known to be unsolvable by computers. That is, no one can write a computer programme to answer that question. It is known to be unsolvable. But, how can we tell it is unsolvable ?. How can we tell that such a programme can not be written? You can not try all possible solution methods and see they all fail. You can not think of all (candidate) methods to solve the Halting Problem. Thus you need some kind of formal approaches here to avoid dealing with a extremely large number (if not infinite) of possibilities.

Discrete mathematics is the foundation for the formal approaches. It discusses languages used in mathematical reasoning, basic concepts, and their properties and relationships among them. Though there is no time to cover them in this course, discrete mathematics is also concerned with techniques to solve certain types of problems such as how to count or enumerate quantities.

The kind of counting problems includes: How many routes exist from point A to point B in a computer network ? How much execution time is required to sort a list of integers in increasing order ? What is the probability of winning a lottery? What is the shortest path from point A to point B in a computer network ?

The subjects covered in this course include propositional logic, predicate logic, sets, relations, and functions, in particular growth of function.

The first subject is logic. It is a language that captures the essence of our reasoning, and correct reasoning must follow the rules of this language. We start with logic of sentences called propositional logic, and study elements of logic, (logical) relationships between propositions, and reasoning.

Then we learn a little more powerful logic called predicate logic. It allows us to reason with statements involving variables among others.

Earlier we also study sets, relations between sets, and operations on sets. Just about everything is described based on sets, when rigor is required. It is the basis of every theory in computer science and mathematics. We learn mathematical reasoning, in particular recursive definitions and mathematical induction. There are sets, operations and functions that can be defined precisely by recursive definitions.

Properties of those recursively defined objects can be established rigorously using proof by induction. They are an abstraction of relations we are familiar with in everyday life such as husband–wife relation, parent–child relation and ownership relation. They are also one of the key concepts in the discussion of many subjects on computer and computation.

For example, a database is viewed as a set of relations and database query languages are constructed based on operations on relations and sets. Graphs are also covered briefly here. They are an example of discrete structures and they are one of the most useful models for computer scientists and engineers in solving problems. Functions are a special type of relation and basically the same kind of concept as the ones we see in calculus.

Before we start the study of discrete structures, we briefly learn general framework of problem solving. If you are a good problem solver, you may skip that and go to logic.

PROBLEM SOLVING

Everyone must have felt at least once in his or her life how wonderful it would be if we could solve a problem at hand preferably without much difficulty or even with some difficulties. Unfortunately the problem solving is an art at this point and there are no universal approaches one can take to solving problems. Basically one must explore possible avenues to a solution one by one until one comes across a right path to a solution. Thus generally speaking, there is guessing and hence an element of luck involved in problem solving. However, in general, as one gains experience in solving problems, one develops one's own techniques and strategies, though they are often intangible. Thus the guessing is not an arbitrary guessing but an educated one.

A FRAMEWORK FOR PROBLEM SOLVING

The following four phases can be identified in the process of solving problems:

1. Understanding the problem
2. Making a plan of solution
3. Carrying out the plan
4. Looking back i.e. verifying

Each of the first two phases is going to be explained below a little more in detail. Phases (3) and (4) should be self–explanatory.

Understanding the Problem

Needless to say that if you do not understand the problem you can never solve it. It is also often true that if you really understand the problem, you can see a solution.

Below are some of the things that can help us understand a problem.

1. **Extract the principal parts of the problem.**

The principal parts are:

For "find" type problems, such as "find the principal and return for a given investment", UNKNOWNS, DATA and CONDITIONS, and for "proof" type problems HYPOTHESIS and CONCLUSION.

1. **Be careful about hidden assumptions, data and conditions.**
2. **Consult definitions for unfamiliar (often even familiar) terminologies.**
3. **Construct one or two simple example to illustrate what the problem says.**

Devising a Solution Plan

Start with the consideration of the principal parts: unknowns, data, and conditions for "find" problems, and hypothesis, and conclusion for "prove" problems. Once you identify the principal parts and understand them, the next thing you can do is to consider the problem from various angles and seek contacts with previously acquired knowledge. The first thing you should do is to try to find facts that are related to the problem at hand. Relevant facts usually involve words that are the same as or similar to those in the given problem. It is also a good idea to try to recall previously solved similar problems.

Look for a helpful idea that shows you the way to the end. Even an incomplete idea should be considered. Go along with it to a new situation, and repeat this process. There is no single method that works for all problems. However, there is a wealth of heuristics we can try. Following are some of the often used heuristics. Add your own heuristics as you gain experience.

1. Have I seen it before ?

That is, do I know similar or related problems ? Similar/ related problems are ones with the same or a similar unknown or unknown may be different but the settings are the same or similar.

1. Do a little analysis on rélationships among data, conditions and unknowns, or between hypothesis and conclusion.
3. What facts do I know related to the problem on hand ?

These are facts on the subjects appearing in the problem. They often involve the same or similar words. It is very important that we know inference rules.

Definitions: Make sure that you know the meaning of technical terms. This is obviously crucial to problem solving at any level. But especially at this level, if you know their meaning and understand the concepts, you can see a solution to most of the problems without much difficulty.

1. Compose a wish list of intermediate goals and try to reach them.
2. Have you used all the conditions/hypotheses ?: When you are looking for paths to a solution or trying to verify your solution, it is often a good idea to check whether or not you have used all the data/hypotheses. If you haven't, something is often amiss.
3. *Divide into cases*: Sometimes if you divide your problem into a number of separate cases based on a property of objects appearing in the problem, it simplifies the problem and clear your mind. For example if the problem concerns integers, then you may want to divide it into two cases: one for even numbers and the other for odd numbers.
4. *Proof by contradiction*: If you make an assumption, and that assumption produces a statement that does not make sense, then you must conclude that your assumption is

wrong. For example, suppose that your car does not start. A number of things could be wrong. Let us assume for simplicity's sake that either the battery is dead or the starter is defective. So you first assume that the battery is dead and try to jump start it. If it doesn't start, you have a situation that does not make sense under your assumption of dead battery. That is, a good battery should start the car but it doesn't. So you conclude that your assumption is wrong. That is the battery is not the cause. Proof by contradiction follows that logic.

In this method we first assume that the assertion to be proven is not true and try to draw a contradiction i.e. something that is always false. If we produce a contradiction, then our assumption is wrong and therefore the assertion we are trying to prove is true. When you are stuck trying to justify some assertion, proof by contradiction is always a good thing to try.

1. Transform/Restate
2. *Working backward*: In this approach, we start from what is required, such as conclusion or final (desired) form of an equation etc., and assume what is sought has been found. Then we ask from what antecedent the desired result could be derived. If the antecedent is found, then we ask from what antecedent that antecedent could be obtained.... We repeat this process until either the data/hypotheses are reached or some easy to solve problem is reached.
3. *Simplify the problem if possible*: Take advantage of symmetries which often exist.

Keep in mind that your first try may not work. But don't get discouraged. If one approach doesn't work, try another. You have to keep trying different approaches, different ideas. As you gain experience, your problem solving skills improve and you tend to find the right approach sooner.

Let us now look at some examples to illustrate the topics discussed above.

Example: This is an example in which you can find a solution once you analyse and understand the unknowns and data.

Problem: A survey of TV viewers shows the following results:

To the question "Do you watch comedies ?", 352 answered "Yes".,

To the question "Do you watch sports ?", 277 answered "Yes", and

To the question "Do you watch both comedies and sports ?", 129 answered "Yes".

Given these data, find, among people who watch at least one of comedies and sports, percentages of people who watch at least one of comedies and sports watch only comedies, only sports, and both comedies and sports.

Let us try to solve this problem following the framework presented above.

Understanding the Problem: This is a "find" type problem. So we try to identify unknowns, data and conditions.

The unknowns are the percentage of people who watch only comedies, the percentage of people who watch only sports, and the percentage of people who watch both comedies and sports. The data are the three numbers: 352, 277 and 129, representing the number of people who watch comedies, sports, and both comedies and sports, respectively. Note that 352 includes people who watch both comedies and sports as well as people who watch only comedies. Similarly for 277.

The conditions are not explicitly given in the problem statement. But one can see that the percentages must add up to 100, and they must be nonnegative.

Devising a Solution Plan: Here we first examine the principal parts in detail.

First let us consider the unknowns in more detail. To calculate the percentage of the people who watch only comedies, for example, we need the number of people who watch at least one of comedies and sports, and the number of people who watch only comedies. Thus actually two unknowns are involved in each of the required percentages, and the real unknowns are the number of people in each of the categories, and the number of people who watch at least one of comedies and sports.

Next let us look at the data. First the number 352 is the number of people who watch comedies.

But that is not necessarily that of the people who watch only comedies. It includes that and the number of people who watch both comedies and sports. Similarly for the second number 277.

Let us use symbols to represent each of the unknowns: Let C represent the number of people who watch only comedies, S that of the people who watch only sports, and T that of the people who watch at least one of those programs.

Then we have the following relationships among the unknowns:

$$C + 129 = 352$$
$$S + 129 = 277$$
$$C + S + 129 = T$$

From these equations we can easily obtain $C = 223$, $S = 148$, and $T = 500$.

Thus the required percentages are 44.6%, 29.6%, and 25.8%, respectively.

All we had to do to solve this problem is to analyse relationships between the data and the unknowns, that is, nothing much beyond "understanding the problem".

Example: This is a problem which you can solve using similar known results.

Problem: Find the (length of) diagonal of a rectangular parallelepiped given its length, width and height.

Again let us try to solve this problem following the framework presented above.

Understanding the Problem: This is a "find" type problem. So we try to identify unknowns, data and conditions.

The unknown is the diagonal of a rectangular parallelepiped, and the data are its length, width and height. Again there are no explicitly stated conditions. But the unknown and data must all be a positive number.

Before proceeding to the next phase, let us make sure that we understand the terminologies.

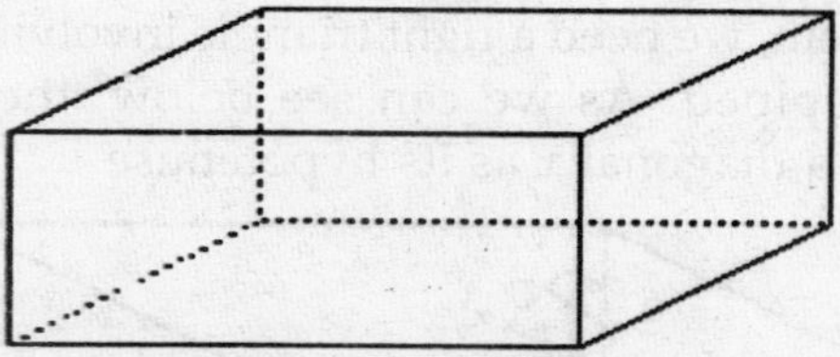

Fig. Rectangular Parallelpiped

First a rectangular parallelepiped is a box with rectangular faces like a cube except that the faces are not necessarily a square but a rectangle as shown below.

Next a diagonal of a rectangular parallelepiped is the line that connects its two vertices (corner points) that are not on the same plane. It is shown in the figure below.

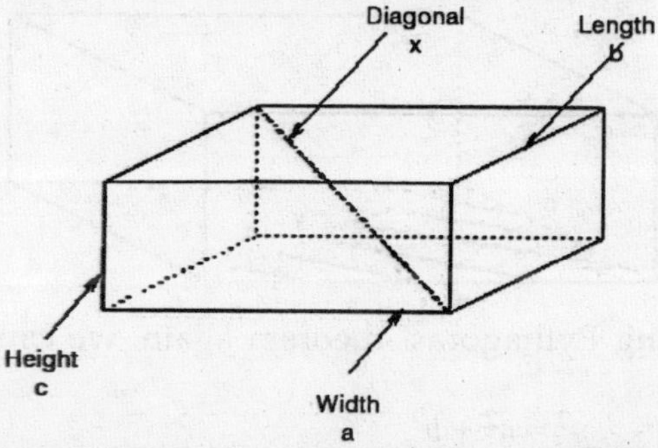

Devising a Solution Plan: Here we first try to find relevant facts. Relevant facts often involve the same or similar words or concepts. Since the unknown is a diagonal, we look for facts concerning diagonal. Note that drawing figures here is quite helpful.

One of the facts that immediately comes to our mind in this problem is Pythagoras' theorem. It has to do with right triangles and is shown below.

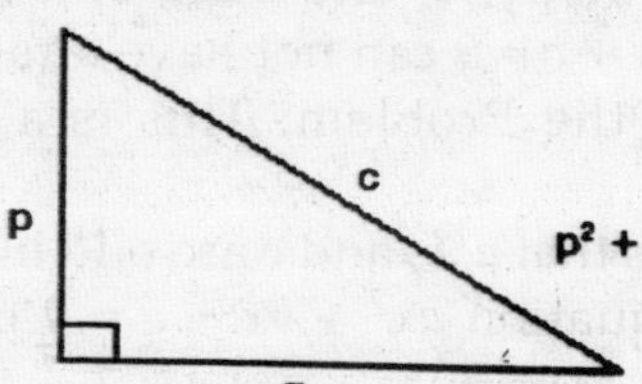

Let us try to see whether or not this theorem helps. To

use this theorem, we need a right triangle involving a diagonal of a parallelepiped. As we can see below, there is a right triangle with a diagonal x as its hypotenuse.

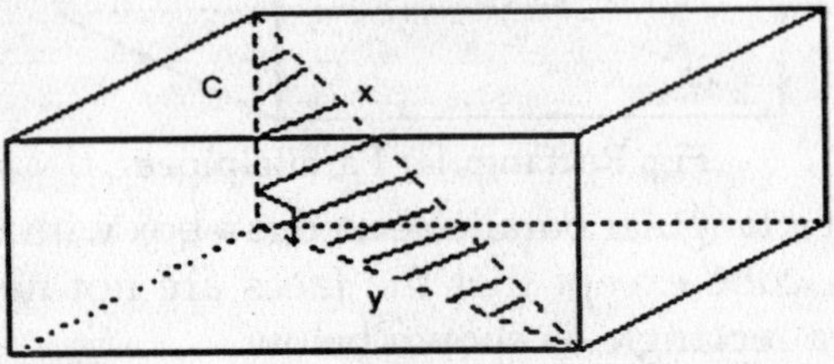

However, the triangle here involves two unknowns: x and y. Since x is what we are looking for, we need to find the value of y. To find y, we note another right triangle shown below.

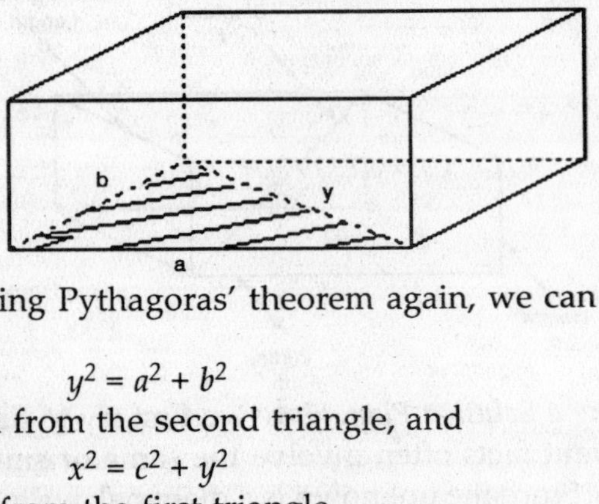

Applying Pythagoras' theorem again, we can obtain the value of y.

Thus $y^2 = a^2 + b^2$

is obtained from the second triangle, and

$$x^2 = c^2 + y^2$$

is derived from the first triangle.

From these two equations, we can find that x is equal to the positive square root of $a^2 + b^2 + c^2$.

Example: This is a proof type problem and "proof by contradiction" is used.

Problem: Given that a, b, and c are odd integers, prove that equation $ax^2 + bx + c = 0$ can not have a rational root.

Understanding the Problem: This is a "prove" type problem.

The hypothesis is that 'a,' b, and c are odd integers, and the conclusion is that equation $ax^2 + bx + c = 0$ can not have a rational root.

The hypothesis is straightforward. In the conclusion,

"rational root" means a root, that is, the value of x that satisfies the equation, and that can be expressed as m/n, where m and n are integers. So the conclusion means that there is no number of the form m/n that satisfies the equation under the hypothesis.

Devising a Solution Plan: For this problem, let us try "proof by contradiction". When you are asked to prove the impossibility of an event or non–existence of certain things, this approach often is quite helpful. Following the "proof by contradiction", let us assume that the conclusion is false, that is the equation $ax^2 + bx + c = 0$ has a rational root m/n, where m and n are integers, when a, b, and c are odd integers. We can assume without loss of generality that m and n do not have any factors in common. Then

$$a(m/n)^2 + b(m/n) + c \quad = 0.$$

Let us try to derive a contradiction from this.

First let us make this equation simpler, that is, let us get rid of fractions.

Since n is not equal to 0, multiplying the both sides of (1) by n^2, we get

$$am^2 + bmn + cn^2 \quad = 0.$$

Since m is an integer, it is either even or odd. We are going to consider those cases one by one. That is "divide into cases".

Let us first consider the case when m is even.

Then n is odd, since otherwise m and n have a common factor 2. Now $am^2 + bmn$ is even, and cn^2 is odd. Hence $am^2 + bmn + cn^2$ can not be 0.

Next let us consider the case when m is odd.

By an argument similar to the previous case, we can see that n is also odd.

If m and n are odd, then am^2, bmn, and cn^2 are all odd, since a, b, and c are odd integers. However, the sum of three odd numbers can not be equal to 0.

Thus by assuming that the conclusion is false, we have arrived at a contradiction, that is m/n does not satisfy the equation. Hence our assumption must be wrong, and therefore the conclusion is correct.

Example: This is another proof type problem and "working backward" technique is used.

Problem: Prove that $(a + b + c)^2 \leq 4(ab + bc + ca)$, if a, b, c are three sides of a triangle.

Understanding the Problem: This is a "prove" type problem.

The hypothesis is that a, b, and c are three sides of a triangle, and the conclusion is that inequality $(a + b + c)^2 \leq 4(ab + bc + ca)$ holds.

Devising a Solution Plan: Here we try "Working Backward" heuristic. That is manipulate the conclusion possibly using the hypothesis and reduce it into something that is obviously true.

First by multiplying out the left hand side of the inequality, $(a + b + c)^2 = a^2 + b^2 + c^2 \leq 2(ab + bc + ca)$.

Hence if $a^2 + b^2 + c^2$ $2(ab + bc + ca)$, then the conclusion holds.

Next, to see what we can try, note that we have not used the hypothesis yet, and see if it can help here.

It is well known that the sum of two sides of a triangle is greater than the third side.

Hence $a + b > c$, $b + c > a$, and $c + a > b$ hold.

From these we can obtain $c(a + b) > c^2$, $a(b + c) > a^2$, and $b(c + a) > b^2$.

By adding these three inequalities, we get

$a^2 + b^2 + c^2 < a(b + c) + b(c + a) + c(a + b) = 2(ab + bc + ca)$.

Hence $a^2 + b^2 + c^2 < 2(ab + bc + ca)$.

Hence $a^2 + b^2 + c^2$ $2(ab + bc + ca)$.

Hence $(a + b + c)^2$ $4(ab + bc + ca)$ holds.

Example: This is a find type problem and "working backward" technique is used.

Problem: Given a 4 quart pail and a 9 quart pail, obtain 6 quarts of water in the 9 quart pail using these two pails. You can fill or empty the pails and you can have as much water as you want.

Understanding the Problem: This is a "find" type problem.

The problem is to obtain 6 quarts of water in the 9 quart pail using 4 quart and 9 quart pails as measures. You can fill either pail from the water source or from the other pail, and you can empty the pails any time.

Devising a Solution Plan: You can solve this in a number of different ways. Here we try "Working Backward" heuristic. It starts with the desired solution and work backward step by step. At each step we try to find a state that immediately precedes the current state such that we can reach the current state from that state with one simple operation such as filling a pail or emptying a pail in this problem. We repeat this process until we reach some easily reachable state such as empty pails, full pails, one pail full and the other empty, etc.

Our solution to the original problem is obtained by traversing this process backward to the desired state.

Let us denote the 9 quart pail by *A* and the 4 quart pail by *B* for simplicity.

In this problem, the desired state is to have 6 qts in *A*.

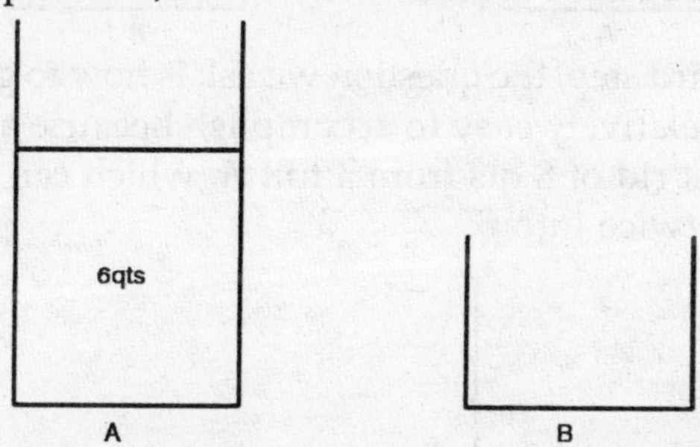

Thus in the first step of "working backward", we ask how we could get to the desired state with one operation. As one can easily see if we could dump 3 qts from 9 qts in *A*, then we would have 6 qts in *A*. To be able to dump 3 qts from *A* we need 1 qt in *B*. Thus the state immediately preceding the current state is the one where *A* is full and *B* has 1 qt in it.

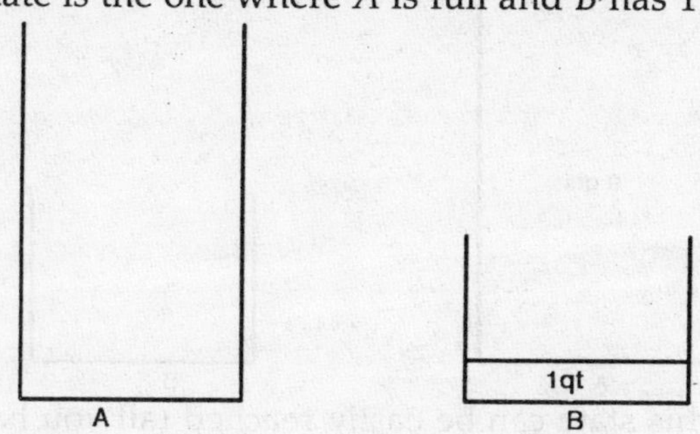

In the second step, the question we ask is how to get 1 qt in *B*.

It does not look easy to get 1 qt in *B*. So let us see whether or not we can get 1 qt in *A*. If we have 1 qt in *A*, then we can certainly get 1 qt in *B* without any trouble. thus we might say that the third state is to have 1 qt in *A*.

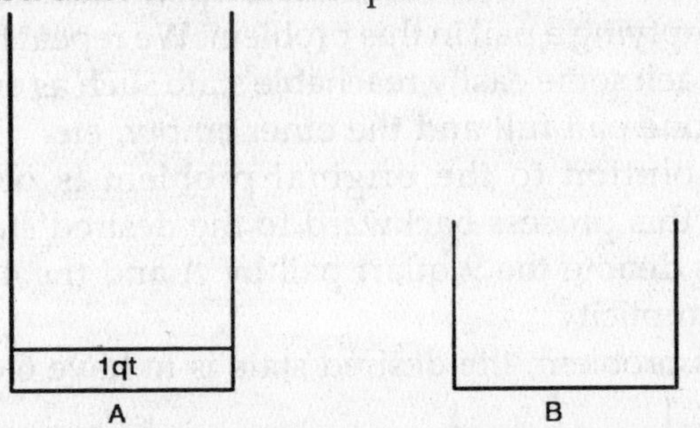

In the third step, the question we ask is how to get 1 qt in *A*.

This is relatively easy to accomplish because all you have to do is to get rid of 8 qts from a full *A*, which can be done by emptying *A* twice into *B*.

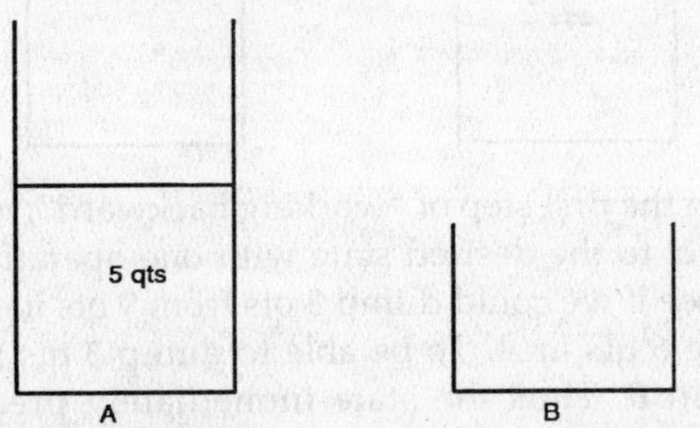

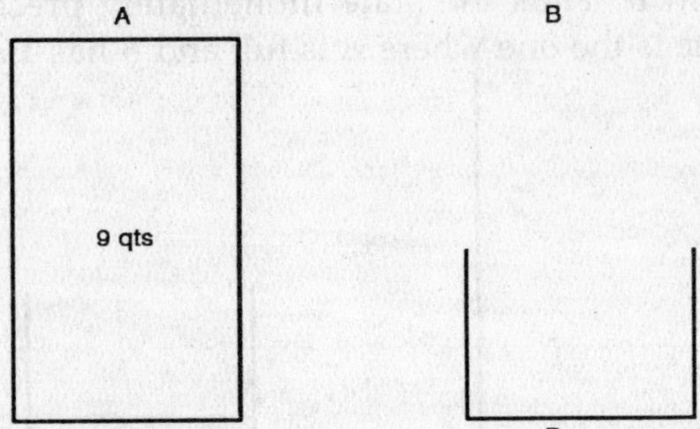

Since this state can be easily reached (all you have to do

to get to this state is to fill A with water), we stop here. Our solution to the original problem is now obtained by going this process backward.

Thus first we fill up A. Then dump A into B leaving 5 qts in A. Then dump A into B again. This gives us 1 qt in A. Pour that into B. Then fill A and empty it into B. We now have 6 qts in A, which is what is required.

EXTRAPOLATION

One of the most important thing, if not the most important, in the university education is the cultivation of the ability to extrapolate. Obviously we can not anticipate all the problems you are going to encounter in the future and prepare you for them. So what we try to do is to teach you the essentials and hope that you can deal with deviations or new problems successfully. Here several examples of extrapolation are going to be given to illustrate what I mean by "extrapolation" and how to deal with new problems with the knowledge and experiences you presently have.

Example:

Problem: A survey of TV viewers shows the following results:

To the question "Do you watch comedies ?", 374 replied "Yes".

To the question "Do you watch sports ?", 360 replied "Yes".

To the question "Do you watch detective stories ?", 350 replied "Yes".

To the question "Do you watch comedies and sports ?", 134 replied "Yes".

To the question "Do you watch comedies and detectives ?", 96 replied "Yes".

To the question "Do you watch detectives and sports ?", 241 replied "Yes".

To the question "Do you watch all three ?", 37 replied "Yes".

Find the percentages of people who watch only comedies, only sports, only detective stories, comedies and sports, comedies and detectives, sports and detectives, and all three. By doing the same kind of analyses on the unknowns and data, one can find these percentages.

The total number of people who watch at least one of these programs is

374 + 360 + 350 – 134 – 96 – 241 + 37 = 650,

because each of 134, 96 and 241 is counted twice into 374 + 360 + 350, and 37 is counted in three times and subtracted out three times in 374 + 360 + 350 – 134 – 96 – 241.

Similarly the number of those watching only comedies is

374 – 134 – 96 + 37 = 181,

the number of those watching only detectives is

350 – 241 – 96 + 37 = 50,

the number of those watching only sports is

360 – 134 – 241 + 37 = 22,

the number of those watching only comedies and detectives is

96 – 37= 59,

the number of those watching only comedies and sports is

350 – 37 = 313,

the number of those watching only detectives and sports is

241 – 37 = 204.

The calculation of the percentages are omitted.

Example: 300 people were surveyed on TV programs they watch and all 300 responded. 75 of them say they watch a regular sports programme, 210 say they watch a regular comedy programme, and 36 say they watch both.

There is a special detective story programme which conflicts with the regular sports and comedy programs. If those who watch regular programs do not want to miss them, how many people can watch the special ?

To answer the question, all you have to do is to find the total number of people who watch at least one of the regular sports and comedy programs. Then 300 minus that number is the answer. Thus this problem is essentially the same as Example,

and the answer is 300 – (75 + 210 – 36) = 51.

Example: 300 people were surveyed. 50 of them watch

sports, 140 watch comedies, and 134 do not watch either of them. Then how many of them watch both comedies and sports ?

Since 300 were surveyed and 134 do not watch either sports or comedies, 166 watch at least one. Hence 50 + 140 – 166 = 24 watch both. Thus 50 – 24 = 26 watch only sports, and 140 – 24 = 116 watch only comedies.

Chapter 2

Set Theory

BACKGROUND

The concept of set is fundamental to mathematics and computer science. Everything mathematical starts with sets. For example, relationships between two objects are represented as a set of ordered pairs of objects, the concept of ordered pair is defined using sets, natural numbers, which are the basis of other numbers, are also defined using sets, the concept of function, being a special type of relation, is based on sets, and graphs and digraphs consisting of lines and points are described as an ordered pair of sets.

Though the concept of set is fundamental to mathematics, it is not defined rigorously here. Instead we rely on everyone's notion of "Set" as a collection of objects or a container of objects. In that sense "Set" is an undefined concept here. Similarly we say an object "belongs to "or "is a member of" a set without rigorously defining what it means. "An object(element) x belongs to a set A" is symbolically represented by "x A". It is also assumed that sets have certain (obvious) properties usually asssociated with a collection of objects such as the union of sets exists, for any pair of sets there is a set that contains them etc.

This approach to set theory is called "naive set theory "as opposed to more rigorous "axiomatic set theory". It was first developed by the German mathematician Georg Cantor at the end of the 19th century. For more on naive and axiomatic set theories click here which is not required for this course. Though the naive set theory is not rigorous, it is simpler and practically all the results we need can be derived within the naive set theory.

DEFINITION OF SET

A set can be described in a number of different ways. The simplest is to list up all of its members if that is possible. For example {*1, 2, 3*} is the set of three numbers *1, 2,* and *3.* {indicates the beginning of the set, and} its end. Every object between them separated by commas is a member of the set. Thus {{*1, 2*}, {{*3*}, *2*}, *2*}, {*1*}} is the set of the elemcnts {*1, 2*}, {{*3*}, *2*} and {*1*}.

A set can also be described by listing the properties that its members must satisfy. For example, $\{x \mid 1 \leq x \leq 2$ and x is a real number.$\}$ represents the set of real numbers between *1* and *2*, and $\{x \mid x$ is the square of an integer and $x \leq 100\}$ represents the set {*0, 1, 4, 9, 16, 25, 36, 49, 64, 81, 100*}.

A third way to describe a set is to give a procedure to generate the members of the set. The recursive/inductive definition is an example and it is going to be studied later.

In this representation, first, basic elements of the set are presented. Then a method is given to generate elements of the set from known elements of the set. Thirdly a statement is given that excludes undesirable elements (which may be included in the set otherwise) from the set. for example the set of natural numbers *N* can be defined recursively as the set that satisfies the following (1), (2), and (3):

1. $0 \in N$
2. For any number x if $x \in N$, then $x + 1 \in N$.
3. Nothing is in *N* unless it is obtained from (1) and (2).

Following this definition, the set of natural numbers *N* can be obtained as follows:

First by (1), *0* is put into *N*.

Then by (2), since *0* is in *N*, $0 + 1$ $(= 1)$ is in *N*.

Then by (2) again, $1 + 1$ $(= 2)$ is in *N*.

Proceeding in this manner all the natural numbers are put into *N*.

Note that if we don't have (3), *0.5, 1.5, 2.5, ...* can be included in *N*, which is not what we want as the set of natural numbers.

BASICS OF SET

Definition (Equality of sets): Two sets are equal if and only if they have the same elements.

More formally, for any sets A and B, $A = B$ if and only if x $[x \in A \leftrightarrow x \in B]$.

Thus for example $\{1, 2, 3\} = \{3, 2, 1\}$, that is the order of elements does not matter, and $\{1, 2, 3\} = \{3, 2, 1, 1\}$, that is duplications do not make any difference for sets.

Definition (Subset): A set A is a subset of a set B if and only if everything in A isalso in B.

More formally, for any sets A and B, A is a subset of B, and denoted by $A \subseteq B$, if and only if $\forall x\ [x \in A \rightarrow x\ B]$.

If $A \subseteq B$, and $A \neq B$, then A is said to be a proper subset of B and it is denoted by$A \subset B$.

For example $\{1, 2\} \subseteq \{3, 2, 1\}$.

Also $\{1, 2\} \subset \{3, 2, 1\}$.

Cardinality: If a set S has n distinct elements for some natural number n, n is the cardinality (size) of S and S is a finite set.

The cardinality of S is denoted by $|S|$.

For example the cardinality of the set $\{3, 1, 2\}$ is 3.

Empty set: A set which has no elements is called an empty set.

More formally, an empty set, denoted by ØØ, is a set that satisfies the following:

$$\forall x\ x \notin Ø,$$

where $\notin$ means "is not in" or "is not a member of".

Note that Ø and {Ø} are different sets. {Ø} has one element namely in it. So {Ø} is not empty. But Ø has nothing in it.

Definition(Universal set): A set which has all the elements in the universe of discourse is called a universal set.

More formally, a universal set, denoted by U, is a set that satisfies the following:

$\forall x\ x \in U$.

Three subset relationships involving empty set and

universal set are listed below as theorems without proof. Their proofs are found elsewhere.

Note that the set A in the next four theorems are arbitrary. So A can be an empty set or universal set.

Theorem: For an arbitrary set A $A \subseteq U$.

Theorem: For an arbitrary set A ØÍA .

Theorem: For an arbitrary set A $A \subseteq A$.

Definition(Power set): The set of all subsets of a set A is called the power set of A and denoted by 2^A or $p(A)$.

For example for A = {*1*, 2}, $p(A)$ = { Ø, {*1*}, {2}, {*1*, 2}} .

For B = {{*1*, 2}, {{*1*}, 2}, Ø}, $p(B)$ = { Ø, {{*1*, 2}}, {{{*1*}, 2}}, {Ø}, {{*1*, 2},{{*1*}, 2}Ø}, {{*1*, 2},}, {{{*1*}, 2},}, {{*1*, 2}, {{*1*}, 2}, Ø}} .

Also p(Ø) = {Ø} and ({Ø}) = {Ø, {Ø}}.

Theorem: For an arbitrary set A, the number of subsets of A is $2^{|A|}$.

REASONING

Mathematical theories are constructed starting with some fundamental assumptions, called axioms, such as "sets exist" and "objects belong to a set" in the case of naive set theory, then proceeding to defining concepts(definitions) such as "equality of sets", and "subset", and establishing their properties and relationships between them in the form of theorems such as "Two sets are equal if and only if each is a subset of the other", which in turn causes introduction of new concepts and establishment of their properties and relationships. Proofs are the arguments for establishing those properties and relationships. At the bottom level these arguments follow the inference rules of propositional and predicate logic, that is the conclusion of the theorem being proved must be derived from its hypotheses, axioms, definitions, and proven theorems using inference rules.

However, at the bottom level they become tedious and inefficient as one can easily imagine. Thus in actual proofs short–cuts are taken using already proven theorems, using multiple inference rules in one step without explicitly mentioning them individually, omitting "obvious" proofs, and so on.

Finding a proof is in general an art. There is no single method that works for all cases. However, at this level the most important thing to remember is to know and understand definitions of concepts involved. The next important thing to keep in mind is to look up relevant facts and try to use them. Even if you don't see the entire path to the goal, if you move one step forward from where you are, you get a new perspective and it often gives you some good ideas to pursue. Needless to say that you must not forget the inference rules. It is not a bad idea to review "Problem Solving" we studied earlier here.

There are also some well used and often very useful proof techniques such as trivial proof, vacuous proof, direct proof, proof by contradiction, proving the contrapositive, and proof by induction. These are explained below with proofs of the theorems on subset relation as examples

Theorem: $A \subseteq U$.

Proof: By the definition of $\subseteq$, we need to show that $\forall x[x \in A \rightarrow x \in U]$.

For that, what we need is to show that for an arbitrary x, $x \in A \rightarrow x \in U$ holds according to the inference rule "universal generalization".

Since x is an object of the universe of discourse, $x \in U$ is true for any arbitrary object by the Universal Instantiation. Hence $x \in A \rightarrow x \in U$ is true for any arbitrary object x ($p \rightarrow q$ is always true if q is true regardless of what p is). Thus by the Universal Generalization $\forall x[x \in A \rightarrow x \in U]$, that is $A \subseteq U$, by the definition of subset.

We say $p \rightarrow q$ is *trivially true* if q is true, and this kind of proof (i.e. showing q is true for $p \rightarrow q$ without referring to p) is called a trivial proof.

Theorem: $\emptyset \subseteq A$.

Proof: By the definition of $\subseteq$, we need to show that $\forall x[x \in \emptyset \rightarrow x \in A]$. For that, what we need is to show that for an arbitrary x, $x \in \emptyset \rightarrow x \in A$ holds. Then apply the Universal Generalization.

Since $\forall x \notin \emptyset$, for any arbitrary $x, x \notin \emptyset$ is false by the

Universal Instantiation. Hence $x \in \emptyset \rightarrow x \in A$ is true for any arbitrary x ($p \rightarrow q$ is always true if p is false regardless of what q is). Hence by the Universal Generalization $\forall x[x \in \emptyset \rightarrow x \in A]$. Thus $\emptyset \subseteq A$ by the definition of subset.

We say $p \rightarrow q$ is *vacuously true* if p is false, and this kind of proof (i.e. showing p is false for $p \rightarrow q$) is called a vacuous proof.

Proof: Variations for Theorem

Theorem: Like most others, can be proven in a number of other ways. Here we try to prove it in two other ways.

Proof by Contrapositive: In this method, to prove $p \rightarrow q$ we prove its contrapositive $\neg p \rightarrow \neg q$,, instead.

So to prove $x \in \emptyset \rightarrow x \in A$ for an arbitrary x, try to prove $x \notin A \rightarrow x \notin \emptyset$.

In this case since $x \notin \emptyset$ is true for any arbitrary x, $x \notin A \rightarrow x \notin \emptyset$ is trivially true. Hence its contraposiive is also true.

Proof by Contradiction: In this method, to prove p we assume $\neg p$ and derive a contradiction from that. Then since $\neg p$ implies a contradiction, it can not hold true. Hence p must be true.

So to prove $\forall x[x \in \emptyset \rightarrow x \in A]$ we assume that $\neg \forall x[x \in \emptyset \rightarrow x \in A]$. $\neg \forall x[x \in \emptyset \rightarrow x \in A]$

is equivalent to $\exists x \neg [x \in \emptyset \rightarrow x \in A]$.

This in turn is equivalent to $\exists x[x \in \emptyset \wedge x \notin A]$.

However, $\exists x[x \in \emptyset \wedge x \notin A]$ implies $\exists x[x \in \emptyset]$ by formula 4 of the relationships between the connectives and quantifiers from predicate logic and "simplification" from the implications of propositional logic. But $x \notin \emptyset$ for any x, which contradicts $\exists x[x \in \emptyset]$.

Hence $\neg \forall x[x \in \emptyset \rightarrow x \in A]$, can not be true.

Hence, $\emptyset \subseteq A$.

General Proofs

Theorem: $A = B$ iff $A \subseteq B$ and $B \subseteq A$.

Proof: By the definition of $A = B$,

$$A = B \Leftrightarrow \forall x[x\in A \leftrightarrow x\in B]$$
$$\Leftrightarrow \forall x[x\in A \to x\in B] \wedge (x\in B \to x\in A]$$
$$\Leftrightarrow \forall x[x\in A \to x\in B] \wedge \forall x(x\in B \to x\in A]$$

Since $A \subseteq B \Leftrightarrow \forall x[x\in A \to x\in B]$, this means that $A \subseteq B$ and $B \subseteq A$.

Hence, $A = B$ iff $A \subseteq B$ and $B \subseteq A$.

SET OPERATIONS

Sets can be combined in a number of different ways to produce another set. Here four basic operations are introduced and their properties are discussed.

Union: The union of sets *A* and *B*, denoted by $A \cup B$, is the set defined as

$A \cup B = \{x \mid x \in A\ x \in B\}$

Example: If $A = \{1, 2, 3\}$ and $B = \{4, 5\}$, then $A \cup B = \{1, 2, 3, 4, 5\}$.

Example: If $A = \{1, 2, 3\}$ and $B = \{1, 2, 4, 5\}$, then $A \cup B = \{1, 2, 3, 4, 5\}$.

Note that elements are not repeated in a set.

Intersection: The intersection of sets *A* and *B*, denoted by $A \cap B$, is the set defined as

$A\ B = \{x \mid x\ A\ x\ B\}$

Example: If $A = \{1, 2, 3\}$ and $B = \{1, 2, 4, 5\}$, then $A\ B = \{1, 2\}$.

Example: If $A = \{1, 2, 3\}$ and $B = \{4, 5\}$, then $A \cap B = \emptyset$.

Difference: The difference of sets *A* from *B*, denoted by $A - B$, is the set defined as

$$A - B = \{x \mid x \in A\ x \notin B\}$$

Example: If $A = \{1, 2, 3\}$ and $B = \{1, 2, 4, 5\}$, then $A - B = \{3\}$.

Example: If $A = \{1, 2, 3\}$ and $B = \{4, 5\}$, then $A - B = \{1, 2, 3\}$.

Note that in general $A - B \neq B - A$

Complement: For a set *A*, the difference $U - A$, where *U* is the universe, is called the complement of *A* and it is denoted by $\overline{A}$.

Thus $\overline{A}$ is the set of everything that is not in *A*.

The fourth set operation is the Cartesian product We first define an ordered pair and Cartesian product of two sets using it. Then the Cartesian product of multiple sets is defined using the concept of *n*–tuple.

Orderedpair: An ordered pair is a pair of objects with an order associated with them.

If objects are represented by x and y, then we write the ordered pair as $<x, y>$.

Two ordered pairs $<a, b>$ and $<c, d>$ are equal if and only if $a = c$ and $b = d$. For example the ordered pair $<1, 2>$ is not equal to the ordered pair $<2, 1>$.

Cartesian product: The set of all ordered pairs $<a, b>$, where a is an element of A and b is an element of B, is called the Cartesian product of A and B and is denoted by $A \times B$.

Example: Let $A = \{1, 2, 3\}$ and $B = \{a, b\}$. Then

$A \times B = \{<1, a>, <1, b>, <2, a>, <2, b>, <3, a>, <3, b>\}$.

Ordered n–tuple: An ordered *n*–tuple is a set of *n* objects with an order associated with them (rigorous definition to be filled in). If *n* objects are represented by $x_1, x_2, ..., x_n$, then we write the ordered *n*–tuple as $<x_1, x_2, ..., x_n>$.

Cartesian product: Let $A_1, ..., A_n$ be *n* sets. Then the set of all ordered *n*–tuples $<x_1, ..., x_n>$, where $x_i \in A_i$ for all i, $1 \le i \le n$, is called the Cartesian product of $A_1, ..., A_n$, and is denoted by $A_1 \times ... \times A_n$.

Example: Let $A = \{1, 2\}$, $B = \{a, b\}$ and $C = \{5, 6\}$. Then

$A \times B \times C = \{<1, a, 5>, <1, a, 6>, <1, b, 5>, <1, b, 6>, <2, a, 5>, <2, a, 6>, <2, b, 5>, <2, b, 6>\}$.

Definition (equality of n–tuples): Two ordered *n*–tuples $<x_1, ..., x_n>$ and $<y_1, ..., y_n>$ are equal if and only if $x_i = y_i$ for all i, $1 \le i \le n$.

For example the ordered 3–tuple $<1, 2, 3>$ is not equal to the ordered *n*–tuple$<2, 3, 1>$.

PROPERTIES OF SET

Basic properties of set operations are discussed here. 1 – 6 directly correspond to identities and implications of propositional logic, and 7 – 11 also follow immediately from them as illustrated below.

1. $A \cup \emptyset = A$

   $A \cap U = A$

   .... Identity Laws

2. $A \cup U = U$

   $A \cap \emptyset = \emptyset$

   .... Domination Laws

3. $A \cup A = A$

   $A \cap A = A$

   .... Idempotent Laws

4. $A \cup B = B \cup A$

   $A \cap B = B \cap A$

   .... Commutative Laws

5. $(A \cup B) \cup C = A \cup (B \cup C)$

   $(A \cap B) \cap C = A \cap (B \cap C)$

ASSOCIATIVE LAWS

6. $A \cup (B \cap C) = (A \cup B) \cap (A \cup C)$

   $A \cap (B \cup C) = (A \cap B) \cup (A \cap C)$

DISTRIBUTIVE LAWS

7. If $A \subseteq B$ and $C \subseteq D$, then $A \cup C \subseteq B \cup D$, and $A \cap C \subseteq B \cap D$.
8. If $A \subseteq B$, then $A \cup B = B$ and $A \cap B = A$.
9. $A \cup (B - A) = A \cup B$
10. $A \cap (B - A) = \emptyset$
11. $A - (B \cup C) = (A - B) \cap (A - C)$ (cf. $\overline{A \cup B} = \overline{A} \cap \overline{B}$)

    $A - (B \cap C) = (A - B) \cup (A - C)$ (cf. $\overline{A \cap B} = \overline{A} \cup \overline{B}$)

DE MORGAN'S LAWS

12. $B = \overline{A}$ if and $A \cup B = U$ only if and $A \cap B = \emptyset$
13. $\overline{\overline{A}} = A$.

ADDITIONAL PROPERTIES:

14. $A \subseteq A \cup B$
15. $A \cap B \subseteq A$

The properties 1 ~ 6, and 11 can be proven using equivalences of propositional logic. The others can also be proven similarly by going to logic, though they can be proven also using some of these properties (after those properties are proven, needless to say). Let us prove some of these properties.

Proof for: We are going to prove this by showing that every element that is in $A \cup B$ is also in $B \cup A$ and vice versa.

Consider an arbitrary element x. Then by the definition of set union

$x \in A \cup B \Leftrightarrow x \in A \vee x \in B$

$\Leftrightarrow x \in A \vee \in B$ by the commutativity of $\Leftrightarrow x \in B \cup A$

by the definition of set union.

Hence by Universal Generalization, every element is in $A \cup B$ is also in $B \cup A$.

Hence $A \cup B = B \cup A$.

Note here the correspondence of the commutativity of $\cup$ and that of $\vee$ This correspondence holds not just for the commutativity but also for others.

Furthermore a similar correspondence exists between $\cap$ and $\wedge$, and between $\subseteq$ and $\rightarrow$.

Proof for: By the definition of the equality of sets, we need to prove that $\forall x[x \in A \cup (B \cap C)$ if and only if $x \in (A \cup B) \cap (A \cup C)]$

For that, considering the Univesal Generalization rule, we need to show that for an arbitrary element in the universe x, $x \in A \cup (B \cap C)$ if and only if $x \in (A \cup B) \cap (A \cup C)$

Here the *only if* part is going to be proven. The *if* part can be proven similarly.

$x \in A \cup (B \cap C) \Leftrightarrow x \in (A \cup B) \cap (A \cup C)$

by the definition of $\cup$. $\Leftrightarrow x \in A \vee (x \in B \wedge x \in C)$

by the definition of $\cap$. $\Leftrightarrow (x \in A \vee x \in B) \wedge (x \in A \vee x \in C)$

by the distribution from the equivalences of propositional logic. $\Leftrightarrow (x \in A \cup B) \wedge (x \in A \cup C)$

by the definition of $\cup$. $\Leftrightarrow x \in (A \cup B) \cap (A \cup C)$
by the definition of $\cap$.

Proof for:

1. If $A \subseteq B$ then $A \cup B = B$.
   Let x be an arbitrary element in the universe.
   Then $x \in A \cup B \Leftrightarrow x \in A \vee x \in B$.
   Since $A \subseteq B$, $x \in A \rightarrow x \in B$.
   Also $x \in B \rightarrow x \in B$.
   Hence $x \in A \cup B \rightarrow x \in B$.
   Hence $A \cup B \subseteq B$.
   Since $B \subseteq A \cup B$ (use "addition" rule), $A \cup B = B$ follows.
2. Similarly for $A \cap B = A$.
   Alternative proof:
   For example, (b) can be proven as follows:
   First $A \cap B \subseteq A$.
   Then since $A \subseteq A$, and $A \subseteq B$, by 7 $A \cap A \subseteq A \cap B$.
   Since $A \cap A = A$ by 3, $A \subseteq A \cap B$.

Proof for: Let x be an arbitrary element in the universe.

Then $[x \in A \cup (B - A)] \Leftrightarrow [x \in A \vee (x \in \wedge\, x \notin A)]$

$\Leftrightarrow [(x \in A \vee x \in B) \wedge (x \in A \vee x \notin A)]$

$\Leftrightarrow [(x \in A \vee x \in B) \wedge \text{True}]$

$\Leftrightarrow [x \in A \vee x \in B]$

Hence $A \cup (B - A) = A \cup B$.

Alternative proof

This can also proven using set properties as follows.

$A\ (B - A) = A \cup (B \cup \bar{A})$ by the definition of $(B - A)$.

$= (A \cup B) \cap (A \cup \bar{A})$ by the distribution.

$= (A \cup B) \cap U$

$= (A \cup B)$ by 1.

Proof for: Suppose $A \cap (B - A) \Leftrightarrow x \in A \wedge B - A$.

Then there is an element x that is in $A \cap (B - A)$, i.e.

$x \in A \cap (B - A) \Leftrightarrow x \in A \wedge x \in B - A$

$\Leftrightarrow x \in A \wedge (x \in B \wedge x \notin A)$

$\Leftrightarrow (x \in A \wedge \notin A) \wedge x \in B)$

$\Leftrightarrow F$

Hence $A \cap (B - A) \neq \emptyset$ does not hold.

Hence $A \cap (B - A) = \emptyset$.

This can also be proven in the similar manner to 9 above.

Proof for: Let x be an arbitrary element in the universe.

$$x \in A - (B \cup C) \Leftrightarrow x \in A \wedge x \notin B \cup C \Leftrightarrow x \in A \wedge \neg (x \in B \vee x \in C)$$

Then

$\Leftrightarrow x \in A \wedge (x \notin B \wedge x \notin C)$

$\Leftrightarrow (x \in A \wedge x \notin B) \wedge (x \in A \wedge x \notin C)$

$\Leftrightarrow x \in A - B \wedge x \in A - C$

$\Leftrightarrow x \in (A - B) \cap (A - C)$

Hence $A - (B \cup B) = (A - B) \cap (A - C)$

Proof for: (a) $A \cup B = U \wedge A \cap B = \emptyset \Rightarrow B = A$?

Try to prove $B \subseteq \overline{A}$ and $\overline{A} \subseteq B$.

Let x be an arbitrary element in the universe.

Then if $x \in B$, then $x \notin A$ since $A \cap B = \emptyset$. Hence $x \in \overline{A}$.

Hence $B \subseteq \overline{A}$.

If $x \in \overline{A}$, then $x \notin A$. Since $x \in A \vee x \in B$ (from $A \cup B = U$), $x \in B$ must hold. Hence $\overline{A} \subseteq B$.

Hence $B = A$.

(b) $B = \overline{A} = \Rightarrow (A \cup B = U \wedge A \cap B = \emptyset)$

Since $B = \overline{A}$, $A \cup B = A \cup (U - A) = A \cup U$ since $A \subseteq U$.

Also $A \cap B = A \cap (U - A) = \emptyset$ by 10 above.

Proof for: Since $A \cup A = A \cup A$, $A \cup A = U$.

Also since $A \cap \overline{A} = \overline{A} \cap A$, $\overline{A} \cap A = \emptyset$.

Hence A satisfies the conditions for the complement of A.
Hence $A = \bar{\bar{A}}$.

RECURSIVE DEFINITION

Sets which have too many elements to list them up, and for which there are no convenient or obvious predicates to specify their elements can often be defined using a recursive definition (also called inductive definition). It essentially gives a procedure to generate the members of the set one by one starting with some subset of its elements.

In this type of definition, first a collection of elements to be included initially in the set is specified. These elements can be viewed as the seeds of the set being defined. Next, the rules to be used to generate elements of the set from elements already known to be in the set (initially the seeds) are given. These rules provide a method to construct the set element by element starting with the seeds. These rules can also be used to test elements for the membership in the set.

A recursive definition of a set always consists of three distinct clauses:

1. The basis clause (or simply basis) of the definition establishes that certain objects are in the set. This part of the definition specifies the "seeds" of the set from which the elements of the set are generated using the methods given in the inductive clause. The set of elements specified here is called basis of the set being defined.
2. The inductive clause (or simply induction) of the definition establishes the ways in which elements of the set can be combined to produce new elements of the set. The inductive clause always asserts that if objects are elements of the set, then they can be combined in certain specified ways to create other objects. Let us call the objects used to create a new object the parents of the new object, and the new object is their child .
3. The extremal clause asserts that unless an object can be shown to be a member of the set by applying the basis and inductive clauses a finite number of times, the object is not a member of the set.

The set you are trying to define recursively is the set that satisfies those three clauses.

There are a number of other ways of expressing the extremal clause that are equivalent to the extremal clause given above.

EXAMPLES OF RECURSIVE

Example: Definition of the Set of Natural Numbers *N*

The set *N* is the set that satisfies the following three clauses:

Basis Clause: $0 \in N$

Inductive Clause: For any element *x* in N, $x + 1$ is in N.

Extremal Clause: Nothing is in unless it is obtained from the Basis and Inductive Clauses.

The basis for this set *N* is {*0*}. The $x + 1$ in the Inductive Clause is the parent of *x*, and *x* is the child of $x + 1$.

Following this definition, the set of natural numbers *N* can be obtained as follows:

First by the Basis Clause, *0* is put into *N*.

Then by the Inductive Clause, since *0* is in *N*, $0 + 1$ (= *1*) is in *N*. *0* is the parent of *1*, and *1* is the child of *0*.

Then by the Inductive Clause again, $1 + 1$ (= 2) is in *N*. *1* is the parent of 2, and 2 is the child of *1*.

Proceeding in this manner all the "natural numbers" are put into *N*.

Note that if we don't have the Extremal Clause, *0.5, 1.5, 2.5*, ... can be included in *N*, which is not what we want as the set of natural numbers.

Example: Definition of the Set of Nonnegative Even Numbers *NE*

The set *NE* is the set that satisfies the following three clauses:

Basis Clause: $0 \in NE$

Inductive Clause: For any element *x* in NE, $x + 2$ is in *NE*.

Extremal Clause: Nothing is in unless it is obtained from the Basis and Inductive Clauses.

Example: Definition of the Set of Even Integers *EI*

The set *EI* is the set that satisfies the following three clauses:

Basis Clause: $0 \in EI$

Inductive Clause: For any element x in EI, $x + 2$, and $x - 2$ are in EI.

Extremal Clause: Nothing is in EI unless it is obtained from the Basis and Inductive Clauses.

Example: Definition of the Set of Strings S over the alphabet $\{a, b\}$ excepting empty string.

This is the set of strings consisting of a's and b's such as *abbab, bbabaa,* etc.

The set S is the set that satisfies the following three clauses:

Basis Clause: $a \in S$, and $b \in S$.

Inductive Clause: For any element x in S, $ax \in S$, and $bx \in S$.

Here ax means the concatenation of a with x.

Extremal Clause: Nothing is in S unless it is obtained from the Basis and Inductive Clauses.

Tips for recursively defining a set:

For the "Basis Clause", try simplest elements in the set such as smallest numbers (0, or 1), simplest expressions, or shortest strings. Then see how other elements can be obtained from them, and generalize that generation process for the "Inductive Clause".

The set of propositions (propositional forms) can also be defined recursively.

GENERALIZED SET OPERATIONS

As we saw earlier, union, intersection and Cartesian product of sets are associative. For example

$(A \cup B) \cup C = A \cup (B \cup C)$

To denote either of these we often use $A \cup B \cup C$.

This can be generalized for the union of any finite number of sets as $A_1 \cup A_2 .... \cup CA_n$, which we write as

$$\cup_{i=1}^{n} A_i$$

This generalized union of sets can be rigorously defined as follows:

Definition ($\cup_{i=1}^{n} A_i$):

Basis Clause: For $n = 1, \cup_{i=1}^{n} A_i = A_1$.

Inductive Clause: $\cup_{i=1}^{n+1} A_i = (\cup_{i=1}^{n} A_i) \cup A_{n+1}$

Similarly the generalized intersection $\cap_{i=1}^{n} A_i$ and generalized Cartesian product $x_{i=1}^{n} A_i$ can be defined.

Based on these definitions, De Morgan's law on set union and intersection can also be generalized as follows:

Theorem: *(Generalized De Morgan)* $\overline{\cup_{i=1}^{n} A_i}$ $= \cap_{i=1}^{n} \overline{A_i}$, and $\overline{\cap_{i=1}^{n} A_i}$ $= \cup_{i=1}^{n} \overline{A_i}$

Proof: These can be proven by induction on n and are left as an exercise.

MORE EXAMPLES

Example: Recursively define the set $S = \{n^2 \mid n$ is a natural number greater than $0\}$, that is $S = \{1, 4, 9, 16, ...\}$.

S is the set that is defined by the following clauses:

Basis Clause: 1 is in S.

Inductive Clause: If x in S, then $x + 2x^{1/2} + 1$ is in S.

Extremal Clause: Nothing is in S unless it is obtained from the Basis and Inductive Clauses.

How do we get $x + 2x^{1/2} + 1$ in the Inductive Clause ?

Since $S = \{1, 4, 9, 16, ...\}$and its elements can be ordered 1, 4, 9, ..., n^2, for the n–th element, which is n^2, the next element is $(n+1)^2$. Hence if we let x be the n–th element, then $x = n^2$. Hence $n = x^{1/2}$. But the element next to x is $(n+1)^2$ which is equal to $n^2 + 2n + 1$. By substituting $n = x^{1/2}$ into that we obtain $x + 2x^{1/2} + 1$.

Example: Recursively define the set $T = \{2^n - 2 \mid n$ is a natural number greater than $0\}$, that is $T = \{0, 2, 6, 14, 30, ...\}$.

T is the set that is defined by the following clauses:

Basis Clause: 0 is in T.

Inductive Clause: If x is in T, then $2x + 2$ is in T.

Extremal Clause: Nothing is in T unless it is obtained from the Basis and Inductive Clauses.

How do we get $2x + 2$ in the Inductive Clause ?

Since $T = \{0, 2, 6, 14, ...\}$ and its elements can be ordered 0, 2, 6, ..., $2^n - 2$, for the n–th element, which is $2^n - 2$, the next element is $2^{n+1} - 2$. Hence if we let x be the n–th element, then $x = 2^n - 2$. Now $2^{n+1} - 2 = 2*2^n - 2 = 2*(2^n - 2) + 2 = 2x + 2$.

Example: Recursively define the set of strings $U = \{a^n bc^n \mid$ n is a natural number$\}$, that is U is the set $\{b, abc, aabcc, ...\}$.

Since the simplest element in U is b, let us select b as the basis element. Then from b we can obtain *abc* by prepending a and appending c to it. Similarly *aabcc* can be obtained from *abc* by prepending a and appending c to it. Similarly for the rest of the elements of the set U. Thus U is the set that satisfy the following clauses:

Basis Clause: b is in U.

Inductive Clause: If x is in U, then axc is in U.

Extremal Clause: Nothing is in U unless it is obtained from the Basis and Inductive Clauses.

Example: Recursively define the set $V = \{x \mid x$ is a natural number and $\lfloor x/2 \rfloor$ is even$\}$, that is $x = \{0, 4, 8, 12, 16, ...\}$. Note that $\lfloor x \rfloor$ denotes the largest integer that is not larger than x. If x is nonnegative as in this example, it is the integer part of x.

Since 0 is the smallest element of x, let us select 0 as the basis element of V. For the inductive clause note that for $\lfloor x \rfloor$ to increase by 2 to the next even number, x needs to grow by 4. Hence we have the following definition for V:

V is the set that is defined by the following clauses:

Basis Clause: 0 is in V.

Inductive Clause: If x is in V, then $x + 1$ is in V.

Extremal Clause: Nothing is in V unless it is obtained from the Basis and Inductive Clauses.

We have defined four sets recursively. However, we are at this point not exactly sure whether or not those definitions correctly define the sets. To verify those definitions, that is, to check whether or not the elements of each set have the specified properties, we can use proof by induction. Let us try to verify the definition of Example.

Verification of Definition of Set S:

To verify the definition of Example we are going to prove by (general) induction that the elements of the set S is in the form n^2, where n is a natural number greater than 0.

In the Basis Step we prove that the element in the basis, that is 1 in this case, has the property. Then in the Inductive Step we prove that if an arbitrary element,say x, has the property, then the element generated from x has the property.

Basis Step: Since $1 = 1^2$, the basis element 1 of the set S is of the form n^2, where n is a natural number greater than 0.

Inductive Step: Let x be an arbitrary element of S and $x = n^2$ for some natural number n. Then the element generated from x by the Inductive Clause of the definition of S is $x + 2x^{1/2} + 1$.

We are trying to show that $x + 2x^{1/2} + 1$ is of the form n^2.

Since $x = n^2$, $x^{1/2} = n$. Hence $x + 2x^{1/2} + 1 = n^2 + 2n + 1 = (n + 1)^2$, which is the square of a natural number $n + 1$.

End of the proof.

Next let us try to verify the definition of Example .

Verification of Definition of set U:

To verify the definition of Example we are going to prove by (general) induction that the elements of the set U is in the form a^nbc^n, where n is a natural number.

In the Basis Step we prove that the element in the basis, that is b in this case, has the property. Then in the Inductive Step we prove that if an arbitrary element, say x, has the property, then the element generated from x has the property.

Basis Step: b is in the form a^nbc^n with $n = 0$.

Inductive Step: Let x be an arbitrary element of U and $x = a^nbc^n$ for some natural number n. Then the element generated from x by the Inductive Clause of the definition of U is axc which is equal to aa^nbc^nc, which in turn is equal to $a^{(n+1)}bc^{(n+1)}$. Hence if an arbitrary element x has the property, then its children also have the property.

RECURSIVE DEFINITION OF FUNCTION

Condition: The domain of the function you wish to define

recursively must be a set defined recursively. How to define function recursively: First the values of the function for the basis elements of the domain are specified.

Then the value of the function at an element, say x, of the domain is defined using its value at the parent(s) of the element x.

A few examples are given below.

They are all on functions from integer to integer except the last one.

Example: The function $f(n) = n!$ for natural numbers n can be defined recursively as follows:

The function f is the function that satisfies the following two clauses:

Basis Clause: $f(0) = 0! = 1$

Inductive Clause: For all natural number n, $f(n+1) = (n+1) f(n)$.

Note that here Extremal Clause is not necessary, because the set of natural numbers can be defined recursively and that has the extremal clause in it. So there is no chance of other elements to come into the function being defined.

Using this definition, $3!$ can be found as follows:

Since $0\,! = 1$, $1\,! = 1 * 0\,! = 1 * 1 = 1$,

Hence $2\,! = 2 * 1\,! = 2 * 1 = 2$.

Hence $3\,! = 3 * 2\,! = 3 * 2 * 1 = 6$.

Example: The function $f(n) = 2n + 1$ for natural numbers n can be defined recursively as follows:

The function f is the function that satisfies the following two clauses:

Basis Clause: $f(0) = 1$

Inductive Clause: For all natural number n, $f(n+1) = f(n) + 2$.

See above for the extremal clause.

Example: The function $f(n) = 2^n$ for natural numbers n can be defined recursively as follows:

The function f is the function that satisfies the following two clauses:

Basis Clause: $f(0) = 1$

Inductive Clause: For all natural number n, $f(n+1) = 2\,f(n)$.

Example: The function L from the set S of strings over {a, b} to the set of natural numbers that gives the length of a string can be defined recursively as follows:

The function L is the function that satisfies the following two clauses:

Basis Clause: For symbols a and b of the alphabet, $L(a) = 1$ and $L(b) = 1$.

Inductive Clause: For any string x and y of S, $L(xy) = L(x) + L(y)$, where xy is the concatenation of strings x and y.

This function L gives the number of a's and b's

RECURSIVE ALGORITHM

A recursive algorithm is an algorithm which calls itself with "smaller (or simpler)" input values, and which obtains the result for the current input by applying simple operations to the returned value for the smaller (or simpler) input. More generally if a problem can be solved utilizing solutions to smaller versions of the same problem, and the smaller versions reduce to easily solvable cases, then one can use a recursive algorithm to solve that problem. For example, the elements of a recursively defined set, or the value of a recursively defined function can be obtained by a recursive algorithm.

If a set or a function is defined recursively, then a recursive algorithm to compute its members or values mirrors the definition. Initial steps of the recursive algorithm correspond to the basis clause of the recursive definition and they identify the basis elements.

They are then followed by steps corresponding to the inductive clause, which reduce the computation for an element of one generation to that of elements of the immediately preceding generation.

In general, recursive computer programs require more memory and computation compared with iterative algorithms, but they are simpler and for many cases a natural way of thinking about the problem.

Example: Algorithm for finding the k–th even natural number

Note here that this can be solved very easily by simply outputting $2*(k - 1)$ for a given k . The purpose here, however, is to illustrate the basic idea of recursion rather than solving the problem.

Algorithm: Even(positive integer *k*)
Input: *k*, a positive integer
Output: *k*–th even natural number (the first even being *0*)
Algorithm:
if $k = 1$, then return *0*;
else return Even(k–1) + 2 .

Here the computation of Even(*k*) is reduced to that of Even for a smaller input value, that is Even(k–1). Even(*k*) eventually becomes Even(*1*) which is *0* by the first line. For example, to compute Even(*3*), Algorithm Even(*k*) is called with $k = 2$. In the computation of Even(2), Algorithm Even(*k*) is called with $k = 1$. Since Even(*1*) = *0*, *0* is returned for the computation of Even(2), and Even(2) = Even(*1*) + 2 = *2* is obtained. This value *2* for Even(2) is now returned to the computation of Even(*3*), and Even(*3*) = Even(2) + 2 = *4* is obtained.

As can be seen by comparing this algorithm with the recursive definition of the set of nonnegative even numbers, the first line of the algorithm corresponds to the basis clause of the definition, and the second line corresponds to the inductive clause.

By way of comparison, let us see how the same problem can be solved by an iterative algorithm.

Algorithm 1–a: Even(positive integer *k*)
Input: *k*, a positive integer
Output: k–th even natural number (the first even being *0*)
Algorithm:

```
int i, even;
i := 1;
even := 0;
while(i < k) {
even := even + 2;
i := i + 1;
}
return even .
```

Example: Algorithm for computing the *k*–th power of 2

Algorithm 2 Power of 2(natural number k)

Input: *k*, a natural number

Output: *k*–th power of 2

Algorithm: If *k* = *0*, then return *1*;

else return 2*Power of 2(*k* – *1*) .

By way of comparison, let us see how the same problem can be solved by an iterative algorithm.

Algorithm 2–a Power of 2(natural number k)

Input: *k*, a natural number

Output: *k*–th power of 2

Algorithm:

```
int i, power;
i := 0;
power := 1;
while(i < k) {
power := power * 2;
i := i + 1;
}
return power
```

The next example does not have any corresponding recursive definition. It shows a recursive way of solving a problem.

Example: Recursive Algorithm for Sequential Search

Algorithm 3 SeqSearch(*L*, *i*, *j*, *x*)

Input: *L* is an array, *i* and *j* are positive integers, *i*≤ *j*, and *x* is the key to be searched for in *L*.

Output: If *x* is in *L* between indexes *i* and *j*, then output its index, else output *0*.

Algorithm:

```
if i ≤ j, then
{
if L(i) = x, then return i ;
else return SeqSearch(L, i+1, j, x)
}
else return 0.
```

Recursive algorithms can also be used to test objects for membership in a set.

Example: Algorithm for testing whether or not a number x is a natural number

Algorithm 4 Natural(a number x)
Input: A number x
Output: "Yes" if x is a natural number, else "No"
Algorithm:
if $x < 0$, then return "No"
else
if $x = 0$, then return "Yes"
else return Natural($x - 1$)

Example: Algorithm for testing whether or not an expression w is a proposition(propositional form)

Algorithm Proposition(a string w)
Input: A string w
Output: "Yes" if w is a proposition, else "No"
Algorithm:
if w is *1*(true), *0*(false), or a propositional variable, then return "Yes"
else if $w = \sim w_1$, then return Proposition(w_1)
else
if ($w = w_1\ w_2$ or $w_1\ w_2$ or $w_1\ w_2$ or $w_1\ w_2$) and
Proposition(w_1) = Yes and Proposition(w_2) = Yes
then return Yes
else return No
end

PROOF BY INDUCTION

MATHEMATICAL INDUCTION FIRST PRINCIPLE

As we have seen in recursion, the set of natural numbers can be defined recursively, and its elements can be generated one by one starting with *0* by adding *1*. Thus the set of natural numbers can be described completely by specifying the basis element (*0*), and the process of generating an element from a known element in the set.

Taking advantage of this, natural numbers can be proven to have certain properties as follows:

First it is proven that the basis element, that is *0*, has the property in question (basis step). You prove that the seeds (the first generation elements) have the property.

Then it is proven that if an arbitrary natural number, denote it by n, has the property in question, then the next element, that is $n + 1$, has that property (inductive step). Here you prove that the property is inherited from one generation (n) to the next generation ($n + 1$).

When these two are proven, then it follows that all the natural numbers have that property. For since *0* has the property by the basis step, the element next to it, which is *1*, has the same property by the inductive step. Then since *1* has the property, the element next to it, which is 2, has the same property again by the inductive step. Proceeding likewise, any natural number can be shown to have the property. This process is somewhat analogous to the knocking over a row of dominos with knocking over the first domino corresponding to the basis step.

More generally mathematical statements involving a natural number n such as $1 + 2 + \ldots + n = n(n + 1)/2$ can be proven by mathematical induction by the same token. To prove that a statement $P(n)$ is true for all natural number $n \leq n_0$, where n_0 is a natural number, we proceed as follows:

Basis Step: Prove that $P(n_0)$ is true.

Induction: Prove that for any integer $k \geq n_0$, if $P(k)$ is true (called induction hypothesis), then $P(k+1)$ is true.

The first principle of mathematical induction states that if the basis step and the inductive step are proven, then $P(n)$ is true for all natural number $k \geq n_0$.

As a first step for proof by induction, it is often a good idea to restate $P(k+1)$ in terms of $P(k)$ so that $P(k)$, which is assumed to be true, can be used.

Example: Prove that for any natural number n, $0 + 1 + \ldots + n = n(n + 1)/2$.

Proof:

Basis Step: If $n = 0$, then $LHS = 0$, and $RHS = 0 * (0 + 1) = 0$. Hence $LHS = RHS$.

Induction: Assume that for an arbitrary natural number n, $0 + 1 + \ldots + n = n(n + 1)/2$ . ... *Induction Hypothesis*

To prove this for $n+1$, first try to express *LHS* for $n+1$ in terms of *LHS* for n, and somehow use the induction hypothesis.

Here let us try

$$LHS \text{ for } n + 1 = 0 + 1 + \ldots + n + (n + 1) = (0 + 1 + \ldots + n) + (n + 1) .$$

Using the induction hypothesis, the last expression can be rewritten as

$$n(n + 1)/2 + (n + 1) .$$

Factoring $(n + 1)$ out, we get

$$(n + 1)(n + 2)/ 2,$$

which is equal to the *RHS* for $n+1$.

Thus *LHS* = *RHS* for $n+1$.

PROGRAMME CORRECTNESS USING INDUCTION

Loops in an algorithm/programme can be proven correct using mathematical induction. In general it involves something called "loop invariant" and it is very difficult to prove the correctness of a loop. Here we are goin to give a few examples to convey the basic idea of correctness proof of loop algorithms.

First consider the following piece of code that computes the square of a natural number:

(We do not compute the square this way but this is just to illustrate the concept of loop invariant and its proof by induction.)

SQUARE FUNCTION: SQ(N)

```
S ← 0
i ← 0
while i < n
S ← S + n
i ← i + 1
return S
```

Let us first see how this code computes the square of a natural number. For example let us compute 3^2 using it.

First S ← 0 and i ← 0 give S = 0 and i = 0 initially.

Since i < 3, the while loop is entered.

$S \leftarrow 0 + 3$

$i \leftarrow 0 + 1$

producing S = 3 and i = 1.

Since i < 3, the while loop is entered the second time.

$S \leftarrow 3 + 3$

$i \leftarrow 1 + 1$

producing S = 6 and i = 2.

Since i < 3, the while loop is entered the third time.

$S \leftarrow 6 + 3$

$i \leftarrow 2 + 1$

producing S = 9 and $i = 3$.

Since $i = 3$, the while loop is not entered any longer, $S = 9$ is returned and the algorithm is terminted.

In general to compute n^2 by this algorithm, n is added n times.

To prove that the algorithm is correct, let us first note that the algorithm stops after a finite numbe of steps. For i increases one by one from 0 and n is a natural number. Thus i eventually becomes equal to n.

Next, to prove that it computes n^2, we show that after going through the loop k times, S = k*n and i = k hold. This statement is called a loop invariant and mathematical induction can be used to prove it.

Proof by induction.

Basis Step: k = 0. When k = 0, that is when the loop is not entered, S = 0 and i = 0. Hence S = k*n and i = k hold.

Induction Hypothesis: For an arbitrary value m of k, S = m * n and i = m hold after going through the loop m times.

Inductive Step: When the loop is entered (m + 1)–st time, S = m*n and i = m at the beginning of the loop. Inside the loop,

$$S \leftarrow m^*n + n$$

$$i \leftarrow i + 1$$

producing S = (m + 1)*n and i = m + 1.

Thus $S = k^*n$ and $i = k$ hold for any natural number k.

Now, when the algorithm stops, $i = n$. Hence the loop will have been entered n times. Thus $S = n^*n = n^2$. Hence the algorithm is correct.

The next example is an algorithm to compute the factorial of a positive integer.

FACTORIAL FUNCTION: FAC(N)

```
i ← 1
F ← 1
while i ⇐ n
    F ← F * i
    i ← i + 1
return F
```

Let us first see how this code computes the factorial of a positive integer. For example let us compute 3 !.

First $i \leftarrow 1$ and $F \leftarrow 1$ give $i = 1$ and $F = 1$ initially.

Since $i < 3$, the while loop is entered.

$F \leftarrow 1 * 1$

$i \leftarrow 1 + 1$

producing F = 1 and i = 2.

Since i < 3, the while loop is entered the second time.

$F \leftarrow 1 * 2$

$i \leftarrow 2 + 1$

producing F = 2 and i = 3.

Since i = 3, the while loop is entered the third time.

$F \leftarrow 2 * 3$

$i \leftarrow 3 + 1$

producing F = 6 and i = 4.

Since i = 4, the while loop is not entered any longer, F = 6 is returned and the algorithm is terminted.

To prove that the algorithm is correct, let us first note that the algorithm stops after a finite numbe of steps. For i increases one by one from 1 and n is a positive integer. Thus i eventually becomes equal to n.

Next, to prove that it computes n !, we show that after going through the loop k times, F = k ! and i = k + 1 hold. This is a loop invariant and again we are going to use mathematical induction to prove it.

Proof by induction.

Basis Step: k = 1. When k = 1, that is when the loop is entered the first time, F = 1 * 1 = 1 and i = 1 + 1 = 2. Since 1! = 1, F = k! and i = k + 1 hold.

Induction Hypothesis: For an arbitrary value m of k, F = m! and i = m + 1 hold after going through the loop m times.

Inductive Step: When the loop is entered (m + 1)–st time, F = m! and i = (m+1) at the beginning of the loop. Inside the loop,

$F \leftarrow m!*(m + 1)$
$i \leftarrow (m + 1) + 1$

producing F = (m + 1)! and i = (m + 1) + 1.

Thus F = k! and i = k + 1 hold for any positive integer k.

Now, when the algorithm stops, i = n + 1. Hence the loop will have been entered n times. Thus F = n! is returned. Hence the algorithm is correct.

MATHEMATICAL INDUCTION-SECOND PRINCIPLE

There is another form of induction over the natural numbers based on the second principle of induction to prove assertions of the form $\forall x\, P(x)$. This form of induction does not require the basis step, and in the inductive step $P(n)$ is proved assuming $P(k)$ holds for all $k < n$. Certain problems can be proven more easily by using the second principle than the first principle because $P(k)$ for all $k < n$ can be used rather than just $P(n - 1)$ to prove $P(n)$.

Formally the second principle of induction states that if $\forall n\, [\forall k\, [k < n \rightarrow P(k)] \rightarrow P(n)]$, then $\forall n\, P(n)$ can be concluded.

Here $\forall k\, [k < n \rightarrow P(k)]$ is the induction hypothesis.

The reason that this principle holds is going to be explained later after a few examples of proof.

Example: Let us prove the following equality using the second principle:

For any natural number n, $1 + 3 + ... + (2n + 1) = (n + 1)^2$.

Proof: Assume that $1 + 3 + ... + (2k + 1) = (k + 1)^2$ holds for all k, $k < n$.

Then $1 + 3 + ... + (2n + 1) = (1 + 3 + ... + (2n - 1)) + (2n + 1)$ $= n^2 + (2n + 1) = (n + 1)^2$ by the induction hypothesis.

Hence by the second principle of induction $1 + 3 + ... + (2n + 1) = (n + 1)^2$ holds for all natural numbers.

Example: Prove that for all positive integer n, $\Sigma_{i=1}^{n} i\,(i!) = (n + 1)! - 1$

Proof: Assume that

$1 * 1! + 2 * 2! + ... + k * k! = (k + 1)! - 1$ for all k, $k < n$.

Then $1 * 1! + 2 * 2! + ... + (n-1) * (n-1)! + n * n!$
$= n! - 1 + n * n!$ by the induction hypothesis.
$= (n + 1)n! - 1$

Hence by the second principle of induction $\Sigma_{i=1}^{n}\ i\ (i!) = (n + 1)! - 1$ holds for all positive integers.

Example: Prove that any positive integer n, $n > 1$, can be written as the product of prime numbers.

Proof: Assume that for all positive integers k, $n > k > 1$, k can be written as the product of prime numbers.

We are going to prove that n can be written as the product of prime numbers.

Since n is an integer, it is either a prime number or not a prime number. If n is a prime number, then it is the product of 1, which is a prime number, and itself. Therefore the statement holds true.

If n is not a prime number, then it is a product of two positive integers, say p and q. Since both p and q are smaller than n, by the induction hypothesis they can be written as the product of prime numbers (Note that this is not possible, or at least very hard, if the First Principle is being used). Hence n can also be written as the product of prime numbers.

Chapter 3

Relation and Functions

DEFINITION

In mathematics, a composite function, formed by the composition of one function on another, represents the application of the former to the result of the application of the latter to the argument of the composite. The functions $f : X \rightarrow Y$ and $g : Y \rightarrow Z$ can be composed by first applying f to an argument x and then applying g to the result. Thus one obtains a function $(g \circ f) : X \rightarrow Z$ defined by $(g \circ f)(x) = g(f(x))$ for all x in X. The notation g o f is read as "g circle f" or "g composed with f".

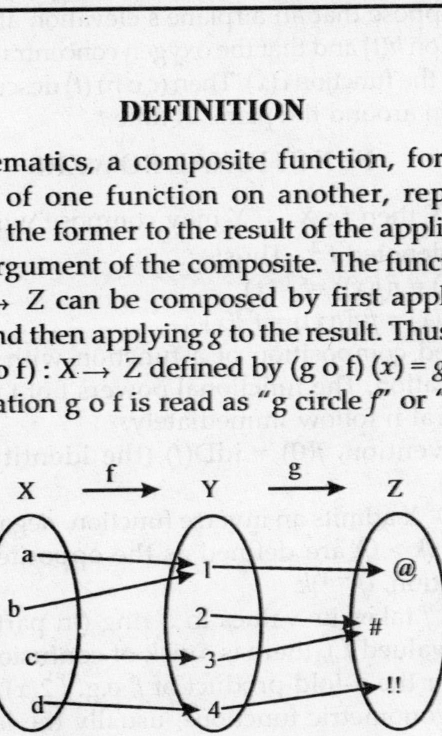

Fig. The Composition of f and g

The composition of functions is always associative. That is, if f, g, and h are three functions with suitably chosen domains and codomains, then $f \circ (g \circ h) = (f \circ g) \circ h$. Since there is no distinction between the choices of placement of parentheses, they may be safely left off.

The functions g and f commute with each other if $g \circ f = f \circ g$. In general, composition of functions will not be commutative. Commutativity is a special property, attained only by particular functions, and often in special circumstances. For example, $|x| + 3 = |x + 3|$ only when $x \geq 0$. But inverse functions always commute to produce the identity mapping.

Derivatives of compositions involving differentiable functions can be found using the chain rule. "Higher" derivatives of such functions are given by Faa di Bruno's formula. As an example, suppose that an airplane's elevation at time t is given by the function $h(t)$ and that the oxygen concentration at elevation x is given by the function $c(x)$. Then (c o h) (t) describes the oxygen concentration around the plane at time t.

FUNCTIONAL POWERS

If $Y \subseteq X$ then $f : X \rightarrow Y$ may compose with itself; this is sometimes denoted f^2. Thus:

(f o f)(x) = $f(f(x))$ = $f^2(x)$

(f o f o)(x) = $f(f(f(x)))$ = $f^3(x)$

Repeated composition of a function with itself is called function iteration. The functional powers f of f n = f n o f = f n + 1 for natural n follow immediately.

By convention, $f(0)$ = idD(f) (the identity map on the domain of f).

If $f : X \rightarrow X$ admits an inverse function, negative functional powers f^{-k} ($k > 0$) are defined as the opposite power of the inverse function, $(f^{-1})k$.

Note: If f takes its values in a ring (in particular for real or complex-valued f), there is a risk of confusion, as $f\,n$ could also stand for the n-fold product of f, e.g. $f\,2(x) = f(x) \,.\, f(x)$.

(For trigonometric functions, usually the latter is meant, at least for positive exponents. For example, in trigonometry, this superscript notation represents standard exponentiation when used with trigonometric functions: sin2(x) = sin(x). sin(x)

However, for negative exponents (especially –1), in nevertheless usually refers to the inverse function, e.g., tan–1 = arctan ($\neq$ 1/tan).

In some cases, an expression for f in $g(x) = f\,r(x)$ can be derived from the rule for g given non-integer values of r. This is called fractional iteration. A simple example would be that where f is the successor function, $f\,r(x) = x + r$.

Iterated functions occur naturally in the study of fractals and dynamical systems.

Composition Mapping

Suppose one has two (or more) functions $f : X \to Y$, $g : X \to X$ having the same domain and range. Then one can form long, potentialy complicated chains of these functions composed together, such as f o f o g o f. Such long chains have the algebraic structure of a maping, sometimes called the composition mapping. In general, composition mappings can have remarkably complicated structure. One particular notable example is the de Rham curve. The set of all functions $f: X \to X$ is called the full transformation semigroup on X.

If the functions are bijective, then the set of all possible combinations of these functions form a group; and one says that the group is generated by these functions.

The set of all bijective functions $f : X \to X$ form a group with respect to the composition operator; this is sometimes called the composition group.

Composition Operator

Given a function g, the composition operator Cg is defined as that operator which maps functions to functions as

$$C_g f = f \circ g$$

Composition operators are studied in the field of operator theory.

Definition: Let $f : S \to T$ and $g : T \to U$ be functions. The composition g o f of f and g is the function from S to U defined by the formula (g o f)(x) = $g(f(x)$ for all $x \in S$.

Proposition: Composition of functions is associative.

Definition: Let f: $S \to T$ be a function then f is said to map S onto T if for each element $y \in T$ there exists an element $x \in S$ with $f(x) = y$.

If $f(x_1) = f(x_2)$ implies $x_1 = x_2$ for all element x_1 $x_2 \in S$, then f is said to be a one-to-one function.

If f is both one-to-one and onto, then it is called a one-to-one correspondence from S to T.

Proposition: Let $f : S \to T$ be a function. Suppose that S and T are finite sets with the same number of elements. Then f is one-to-one if and only if it is onto.

Proposition: Let $f : S \to T$ and $g : T \to U$ be functions.

(*a*) If f and g are one-to-one, the g o f is one-to-one.

(*b*) If f and g are onto, then g o f is onto.

Definition: Let S be a set. The identity function $1_s : S \to S$ is defined by the formula

$$1_s(x) = x \text{ for all } x \in S.$$

If $f : S \to T$ is a function, then a function $g : T \to S$ is called an inverse for f if

$$g \circ f = 1_s \text{ and } f \circ g = 1_T$$

Proposition: Let $f : S \to T$ be a function. If f has an inverse, then it must be one-to-one and onto. Conversely, if f is one-to-one and onto, then it has a unique inverse.

Examples: If the graph of a function is known, how can you decide whether a function is one-to-one (injective) or onto (surjective)

Solution:

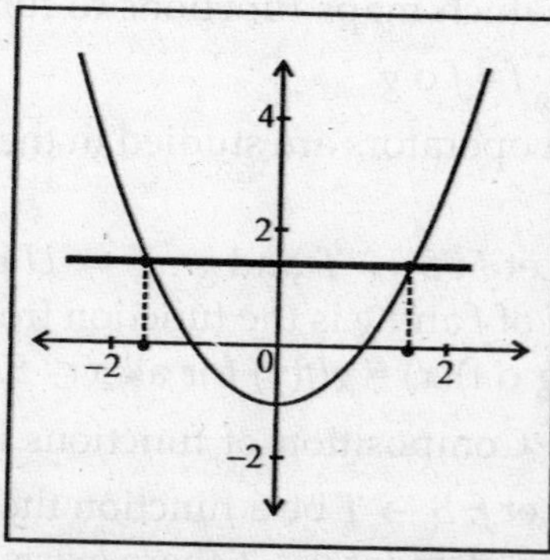

Not a one-to-one Function

Injection: If a horizontal line intersects the graph of a function in more than one place, then there are two different points *a* and *b* for which $f(a) = f(b)$, but $a \# b$.

Then the function is not one-to-one.

Surjection: If you place a light on the left and on the right hand side of the coordinate system, then the shadow of the graph on the y axis is the image of the domain of the function. If that shadow covers the range of the function, then the function is onto.

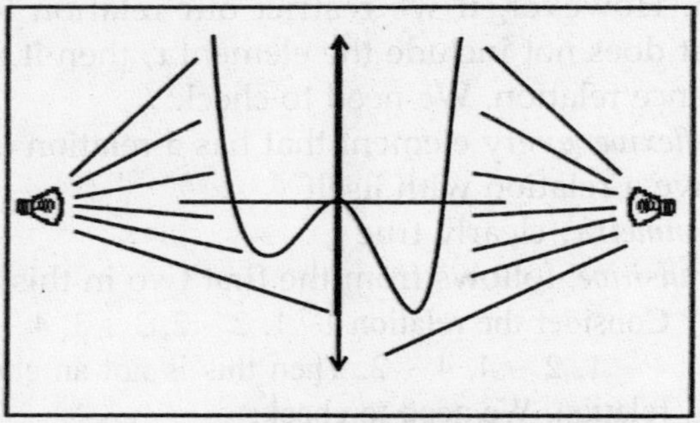

Not an onto Function (if Range is R)

Note that every function can be modified to be onto by setting its range to be the image of its domain.

RELATIONS

Just as there were different classes of functions (bijections, injections, and surjections), there are also special classes of relations. One of the most useful kind of relation (besides functions, which of course are also relations) are those called equivalence relations.

Definition: Equivalence Relation

1. Let S be a set and r a relation between S and itself. We call r and equivalence relation on S if r has the following three properties:

   (*a*) Reflexivity: Every element of S is related to itself

   (*b*) Symmetry: If s is related to t then t is related to s

   (*c*) Transitivity: If s is related to t and t is related to u, then s is related to u.

Examples: Let $A = \{1, 2, 3, 4\}$ and $B = \{a, b, c\}$ and define the following two relations:

1. $r : \{(a, a), (b, b), (a, b), (b, a)\}$
   $s : 1 \sim 1, 2 \sim 2, 3 \sim 3, 4 \sim 4, 1 \sim 4, 4 \sim 1, 2 \sim 4, 4 \sim 2$
3. Which one is an equivalence relation, if any?

Solution: Consider the relation r: $\{(a, a), (b, b), (a, b), (b, a)\}$. Then this is, technically speaking, not an equivalence relation, for the simple reason that the element c has no association. However, if we restrict our relation to a new domain that does not include the element c, then it might be an equivalence relation. We need to check:

1. *Reflexive:* every element that has a relation does also have a relation with itself
2. *Symmetric:* clearly true
3. *Transitive:* follows from the first two in this case.
   (*a*) Consider the relation $1 \sim 1, 2 \sim 2, 3 \sim 3, 4 \sim 4, 1 \sim 4, 4 \sim 1, 2 \sim 4, 4 \sim 2$. Then this is not an equivalence relation. We need to check:

   (*b*) *Reflexive:* every element with a relation is related to itself.

   (*c*) *Symmetric:* true as well (3 is only related to 3, so symmetry condition does not apply here)
3. *Not transitive:* $1 \sim 4$ and $4 \sim 2$, but 1 is not related to 2. At first glance equivalence relations seem to be too abstract to be useful. However, just the opposite is the case. Because they are defined in an abstract fashion, equivalent relations can be utilized in many different situations. In fact, they can be used to define such basic objects as the integers, the rational numbers, and the real numbers. The main result about an equivalence relation on a set A is that it induces a partition of A into disjoint sets. This property is the one that will allow us to define new mathematical objects based on old ones in the next section.

Theorem: Equivalence Classes

1. Let r be an equivalence relation on a set A. Then A can be written as a union of disjoint sets $A = \bigcup_a A_a$ with the following properties:

2. If *a*, *b* are in *A* then $a \sim b$ if and only if *a* and *b* are in the same set A_a
3. 'The subsets A_a are non-empty and pairwise disjoint.
4. The sets *Aa* are called equivalence classes.

Proof: We have to decide what the equivalence classes *Aa* should be: Since by property (1) two elements a and b are supposed to be related if and only if they are in the same class, it seems natural to define:

- $A_a = \{b \in A : b \sim a\}$

To simplify notation a little, we denote this class by *A*(*a*) Now we need to check whether this is a good definition, and whether it satisfies the above properties.

Because of reflexivity of the equivalence relation, the class *A*(*a*) contains the element a and is therefore not empty.

Next, we will show that two classes that are different can bot have any elements in common, i.e. they are disjoint. Recall that disjoint means that two sets have either an empty intersection or are the same sets. Take two elements *a* and *a'* and suppose that *A*(*a*) and *A*(*a'*) have a non-empty intersection; say both classes contain the element *c*. Then

- $c \sim a$, because *c* is in *A*(*a*)
- $a \sim c$, by symmetry and the above line
- $c \sim a$ because *c* is in *A*(*a'*)
- Therefore, by transitivity, $a \sim a'$

Now take any element *b* in *A*(*a'*). Then $b \sim a'$ and $a' \sim a$ so that $b \sim a$ by transitivity, and hence be is contained in *A*(*a*). Therefore *A*(*a'*) is a subset of *A*(*a*). The same argument works the other way around as well (by symmetry), showing that *A*(*a*) is a subset of *A*(*a'*). Hence, *A*(*a*) = *A*(*a'*). This shows that two different classes must be disjoint.

Finally, we need to show that all classes *A*(*a*) make up the original set *A*. But that is clear, because every element a in A belongs to the class *A*(*a*), hence the union of all classes *A*(*a*) must be contained in *A*, and therefore must be equal to *A*.

Example: Let $f : S -> T$ ..... be any function. For $x_1, x_2 \in S$ we define $x_1 \sim x_2$ if $f(x_1) = f(x_2)$. Then for all $x_1, x_2, x_3 \in S$ we have

1. $f(x_1) = f(x_1)$;
2. if $f(x_1) = f(x_2)$, then $f(x_2) = f(x_1)$;
3. if $f(x_1) = f(x_2) = f(x_3)$, then $f(x_i) = (x_3)$.

This shows that we have defined an equivalence relation on the set *S*. The proof of this is easy because the equivalence relation is defined in terms of equality of the images $f(x)$, and equality is the most elementary equivalence relation.

The collection of all equivalence classes of *S* under ~will be denoted by *S*/*f*.

Proposition: Let *S* be a set, and let ~ be an equivalence relation on *S*. Then each element of **S** belongs to exactly one of the equivalence classes of *S* determined by the relation~

Definition: Let *S* be any set. A family *P* of subsets of *S* is called a partition of *S* if each element of *S* belongs to exactly one of the members of *P*.

Proposition: Any partition *P* of a set *S* determines an equivalence relation.

Theorem: If $F: S \rightarrow T$ is any function, and ~ is the equivalence relation defined on *S* by letting

$x_1 \sim x_2$ if $f(x_1) = f(x_2)$ for all x_1, x_2 *S*, then there is a one-to-one correspondence between the elements of the image $f(S)$ of *S* under *f* and the equivalence classes *S*/*f* of the relation ~.

If $f: S \rightarrow T$ is a function and *y* belongs to the image $f(S)$, then the inverse image of *y* is

$$f^{-1}(y) = \{x \in S \mid f(x) = y\}.$$

The inverse images of elements of $f(S)$ are the equivalence classes in *S*/*f*. (Note carefully that we are not implying that *f* has an inverse function.)

Permutations

Definition: Let *S* be a set. A function $\sigma : S -> S$ is called a permutation of *S* if σ is one-to-one and onto.

The set of all permutations of *S* will be denoted by Sym(S).

The set of all permutations of the set {1, 2, ..., *n*} will be denoted by S_n.

Propositio shows that the composition of two permutations in Sym(S) is again a permutation. It is obvious that the identity function on *S* is one-to-one and onto.

Propositions shows that any permutation in Sym(s) has an inverse function that is also one-to-one and onto. We can summarize these important properties as follows:

(i) If $\sigma, \tau \in$ Sym(S), then $\sigma\tau \in$ Sym(S);

(ii) $1_s \in$ Sym(S);

(iii) if, $\sigma \in$ Sym(S) then $\sigma^{-1} \in$ Sym(S).

Definition: Let S be a set, and let $\sigma \in$ Sym(S). Then is called a cycle of length k if there exist elements $a_1, a_2, \ldots\ldots, a_k \in S$ such that

$\sigma(a_1) = a_2$, $\sigma(a_2) = a_3, \ldots\ldots$, $\sigma(a_{k-1}) = a_k$, $\sigma(a_k) = a_1$, and

$\sigma(x) = x$ for all other elements $x \in S$ with $x \neq a_1$ for $i = 1, 2, \ldots\ldots, 1, 2, \ldots\ldots, k$.

In this case we write $\sigma = (a_1, a_2, \ldots\ldots a_k)$.

We can also write $\sigma = (a_1, a_2, \ldots\ldots a_k)$ or $= (a_3, a_k, a_1, a_2)$, etc. The notation for a cycle of length k can thus be written in k different ways, depending on the starting point. The notation (1) is used for the identity permutation.

Definition: Let $\sigma = (a_1, a_2, \ldots\ldots a_k)$ and $\tau = (b_1, a_2, \ldots\ldots b_m)$ be cycles in Sym(S), for a set S. Then σ and τ are said to be disjoint if $a_i \neq b_j$ for all i, j.

Proposition: Let S be any set. If σ and τ are disjoint cycle in Sym (S), then $\sigma\tau = \tau\sigma$.

Theorem: Every permutation in S_n can be written as a product of disjoint cycles. The cycles that appear in the product are unique.

Definition: Let $\sigma \in S_n$ The least positive integer m such that $\sigma^m = (1)$ is called the order of σ.

Proposition: Let $\sigma \in S_n$ have order m. Then for all integers j, k we have

$$\sigma^j = \sigma^k \text{ if and only if } j \equiv k \pmod{m}.$$

Proposition: Let $\sigma \in S_n$ be written as a product of disjoint cycles. Then the order of is the least common multiple of the lengths of its cycle.

Definition: A cycle (a_1, a_2) of length two is called a transposition.

Proposition: Any permutation in S_n where $n \geq 2$ can be written as a product of transpositions.

Theorem: If a permutation is written as a product of transpositions in two ways, then the number of transpositions is either even in both cases or odd in both cases.

Definition: A permutation σ is called even if it can be written as a product of an even number of transpositions, and odd if it can be written as a product of an odd number of transpositions.

CONCEPT OF RELATION

After knowing some of the basic elements of set theory (sets), we will move on to the second most elementary concept, the concept of relations and functions. Let A and B two sets. A relations between A and B is a collection of ordered pairs (a, b) such that $a \in A$ and $b \in B$. Often we use the notation $a \sim b$ to indicated that a and b are related, rather then the order pair notation (a, b). Note that this does not mean that *each* element from A needs to be associated with one (or more) elements from B. It is sufficient if *some* associations between elements of A and B are defined. In contrast, there is the definition of a function:

Thus, a set of ordered pairs is called a relation.

For example, (1, 2), (3, 5), (7, 10) … are relations.

The first components in the ordered pairs (x-coordinate) is called the domain, and the second components (y-coordinate of $f(x)$ coordinate) is called the range.

Functions: Domain and Range

Let's return to the subject of domain and range. When functions are first introduced, you will probably have some slightly pathetic "functions" and relations to deal with, being just sets of points. These won't be terribly useful or interesting functions and relations, but your text wants you to get the idea of what the domain and range are.

The first components in the ordered pairs (x-coordinate) is called the domain, and the second components (y-coordinate or $f(x)$ coordinate) is called the range.

Definition: Let A and B be two sets. Since, a function f from A to B is a relation between A and B such that for each

$a \in A$ there is one and only one associated $b \in B$. The set A is called the domain of the function, B is called its range. Often a function is denoted as $y = f(x)$ or simply $f(x)$ or simply $f(x)$, indicating the relation $\{(x, f(x)\}$.

Definition: Let S and T be sets. A function from S into T is a subset F of $S \times T$ such that each element $x \in S$ there is exactly one element $y \in T$ such the $(x, y) \in F$. The set S is called the domain of the function, and the set T is called the codomain of the function. The subset $\{y \in T \mid (x, y) \in F$ for some $x \in S\}$ of the codomain is called the image of the function.

Functions Versus Relations

There are different ways of looking at functions. We will consider a few. But first, we need to discuss some terminology.

A "relations" is just a relationship between sets of information. Think of all the people in one of your classes, and think of their heights. The pairing of names and heights is a relation. In relations and functions, the pairs of names and heights are "ordered", which means one comes first and the other comes second. To put it another way, we could set up this pairing so that either you give me a name, and then I give you that person's height, or else you give me a height, and I give you the names of all the people who are that tall. (By the way, the set of all the starting points is called "the domain" and the set of all the ending points is called "the range." The domain is where you start; the range is where you go. The domain is the x' s; the range is the y' s. I'll explain more on the subject of determining domains and ranges later.)

A function is a well-behaved relation. Just as with members of your own family, some members of the relation family are better behaved than other. (This means that, while all functions are relations, since they pair information, not all relations are functions. Functions are a sub-classification of relations.) When we say that a function is "a well-behaved relation", we mean that, given a starting point, we know exactly where to go. That is, given an x, we get only and exactly one y.

Let's return to our relation of your classmates and their heights, and let's suppose that the domain is the set of everybody's heights. And let's suppose that there's a pizza-delivery guy waiting in the hallway. And all the delivery guy knows is that the pizza is for the student in your classroom who is five-foot-five. Now let the guy in. Who does he go to? What if you are five-five? And what if you're out of cash? You see, the relation "height indicates name" is not well-behaved. It is not a function. Given the relationship (x, y) = (five-foot-five person, name), there might be six different possibilities for y = "name". For a relation to be a function, there must be only and exactly one y that corresponds to a given x. Here are some pictures of this:

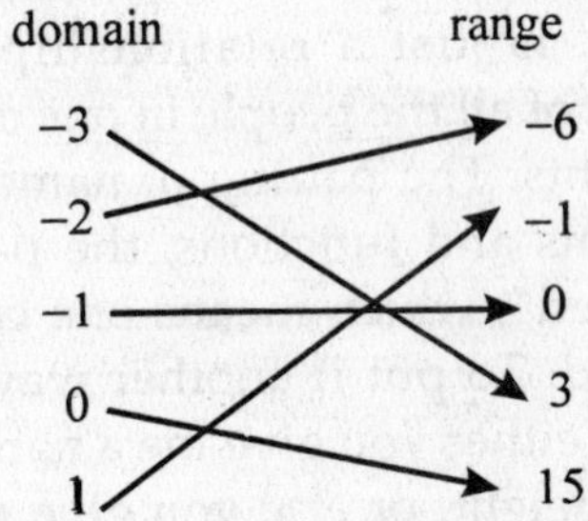

This is a function. You can tell by tracing from each x to each y. There is only one y for each x; there is only one arrow coming from each x.

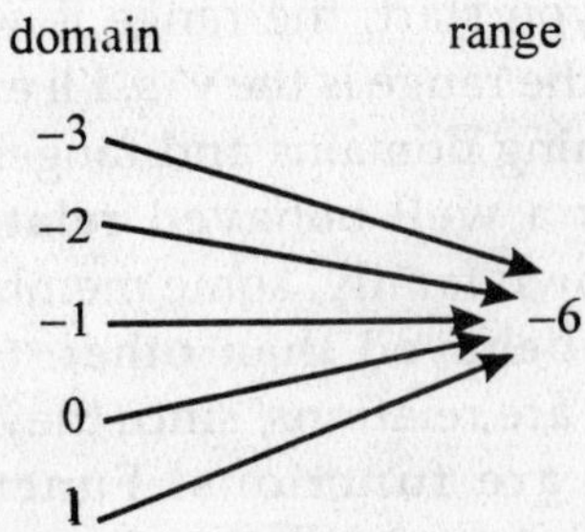

This is a function, There is only one arrow coming from each x; there is only one y for each x. It just so happens that it's always the same y, but it is only one y. So this is a function; it's just an extremely boring function!

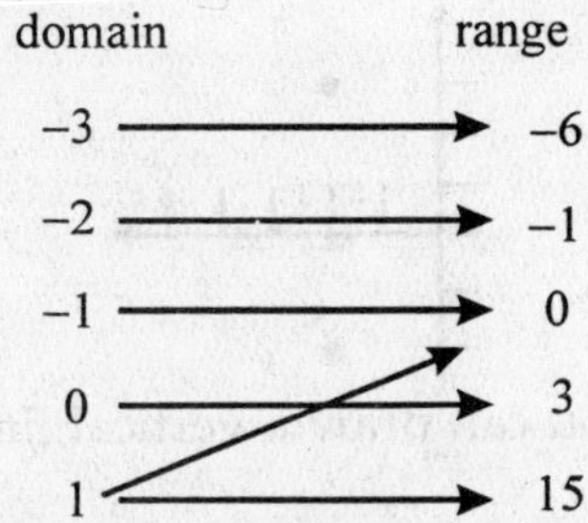

This one is not a function. If you'll notice, there are two arrow coming from the number 1; the number 1 is associated with two different range elements. So this is a relation, but it is not a function.

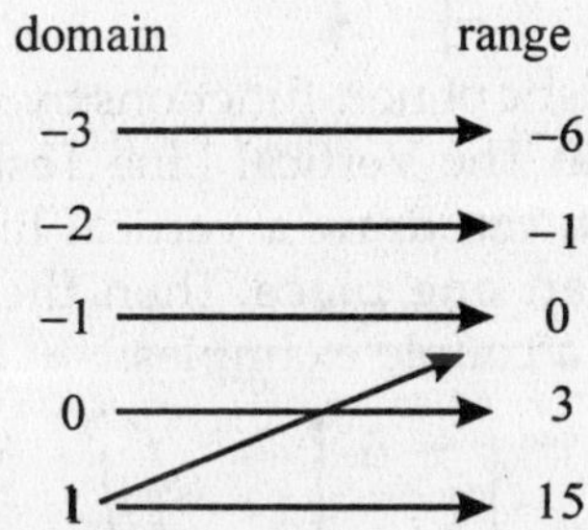

Each element of the domain that has a pair in the range is nicely well-behaved. But what about that 16? It is in the domain, but it has no range element that corresponds to it! This won't work. So then this is not a function.

Vertical Line Test

Given a graph, it can be determined if it is a function or a relation.

In order for the graph to be a function, the vertical line must only intersect the graph at one and only one point.

Looking at this function stuff graphically, what if we had the relation $\{(2, 3),(2, -2)\}$?

We already know that this is not a function, since $x = 2$ goes to both $y = 3$ and $y = -2$.

If we graph this relation, it looks like:

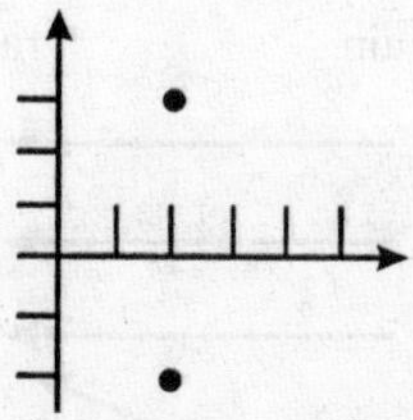

Notice that you can draw a vertical line through the two points, like this:

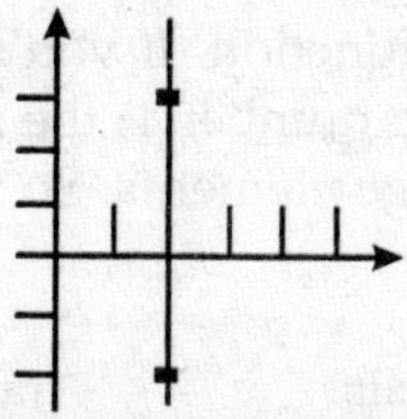

This characteristic of non-functions by I-don't-know-who, and was codified in The Vertical Line Test: Given the graph of a relation, if you can draw a vertical line that crosses the graph in more than one place, then the relation is not a function. Here are a couple examples:

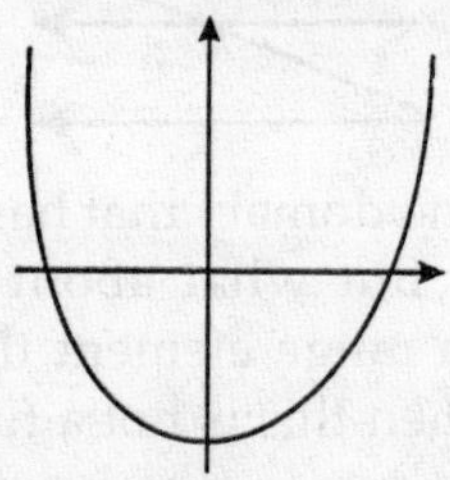

This one is a function. There is no vertical line that will cross this graph twice.

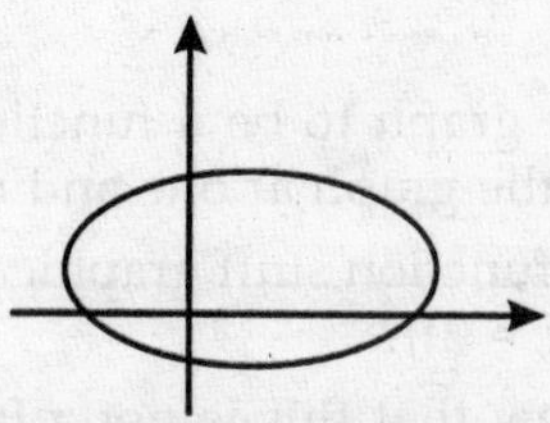

This one is not a function. Any number of vertical lines will intersect this intersect this oval twice. For instance, the y-axis intersects twice.

"Is it a function?" – Quick answer without the graph.

Think of all the graphing that you've done so far. The simplest method is to solve for y. make a T-chart, pick some values for x, solve for the corresponding values of y, plot your points, and connect the dots. Not only is this useful for graphing, but this methodology gives yet another way of identifying functions: If you can solve for "y". then it's a function.

For example, $2y + 3x = 6$ is a function, because you can solve for y:

$$2y + 3x = 6$$
$$2y = -3x + 6$$
$$y = (-3/2)x + 3$$

On the other hand, $y^2 + 3x = 6$ is not a function, because you can not solve for a unique y:

$$y^2 + 3x = 6$$
$$y^2 = -3x + 6$$
$$y = \pm\sqrt{-3x+6}$$

This is solved for "y", but it's not unique. Do you take the positive square root, or the negative? So, in this case, the relation is not a function. (You can also check this by using our first definition from above.

Think of "$x = -1$". Then we get $y2 - 3 = 6$, so $y2 = 9$, and then y can be either –3 or + 3.

That is, if we did an arrow chart, there would be two arrows coming from $x = -1$.)

Thus, a function is a relation were each member of the domain is reserved one and only one member of the range. Functions are named by suing single letters.

For example, $f = \{(x, y) \mid y = x + 4\}$ or $f(x) = x + 4$

The equations above are representations of functions. The second example is the most commonly used representation; it is called the function notation.

The ordered pair for functions is $(x, f(x))$.

Consider the example; $f(x) = x + 4$ at $x = 2$.

$f(2) = 2 + 4 = 6$, thus the ordered pair is (2, 6)

Example: Find the domain and range of: $f(x) = 2/(x-3)$

Solution: Since x cannot be zero; the domain is $D = \{x \mid x^1$ $3\}$ and $f(x)$ can have any value; thus the range is $R = \{f(x) \mid$ any real number$\}$

Example: State the domain and range of the following relation. Is the relation a function?

$$\{(2, -3), (4, 6), (3, -1), (6, 6), (2, 3)\}$$

This list of points, being a relationship between certain x's and certain y's, is a relation. The domain is all the x-values, and the range is all the y-values. you list the values without duplication:

Domain: {2, 3, 4, 6}

Range: {–3, –1, 3, 6}

(It is customary to list these values in numerical order, but it is not required. Sets are called "unordered lists", so you can list the numbers in any order you feel like. Just don't duplicate.)

While this is a relation (because x's and y's are being related to each other), you have two points with the same x-values: (2, –3) and (2, 3). Since $x = 2$ gives you two possible destinations, then this relation is not a function.

Note that all I had to do to check whether the relation was a function was to look for duplicate x- values. If you find a duplicate x-value, then the different y-values mean that you do not have a function.

Example: State the domain and range of the following relation. Is the relation a function?

$$\{(-3, 5), (-2, 5), (-1, 5), (0, 5), (1, 5), (2, 5)\}$$

List the x-values for the domain and the y-values for the range:

Domain: {–3, –2, –1, 0, 1, 2}

Range: {5}

This is another example of a "boring" function (like we saw in the first chart higher up in this page), because all the x-values go to the exact same y-values. But each x-value is different, so, while boring, this relation is indeed a function. (In point of fact, these points lie on the horizontal line $y = 5$.)

There is one other case for finding the domain and range of functions. They will give you a function and ask you to find the domain (and maybe the range, too). I have only ever seen (or can even think of) two things at this stage in your mathematical career that you'll have to check in order to determine the domain of the function they'll give you, and those two things are denominators and square roots.

Example: Determine the domain and range of the given function

$$\frac{x^2 + x - 2}{x^2 - x - 2}$$

Solution: The domain is all the values that x is allowed to take on. The only problem you have with this function is that you have to be careful not to divide by zero. So the only values that x can not take on are those which would cause division by zero. So set the denominator equal to zero and solve; the domain will be everything else.

$$x^2 - x - 2 = 0$$

$$(x - 2)\,(x + 1) = 0$$

$$x = 2 \text{ or } x = -1$$

Then the domain is "all x not equal to – 1 or 2".

The range is a bit trickier, which is why they may not ask for it. In general, though, they'll want you to graph the function and find the range from the picture. In this case:

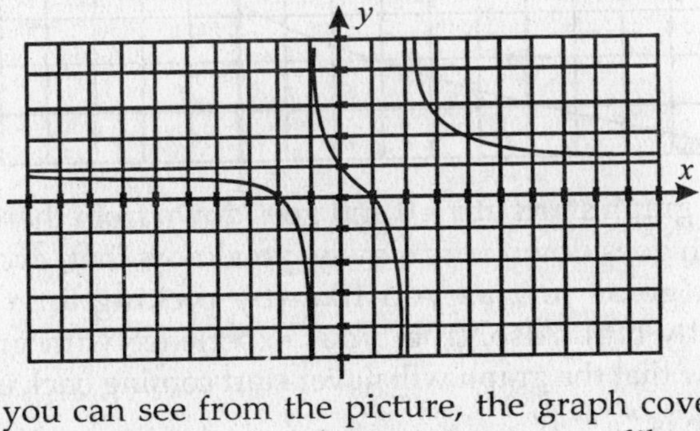

As you can see from the picture, the graph covers all y-values (that is, the graph will go as low as you like, and will also go as high as you like). Since the graph will eventually cover all possible values of y, then the range is "all real numbers".

Example: Determine the domain and range of the given function:

$$-\sqrt{-2x+3}$$

The domain is all values that x can take on. The only problem you have with this function is that you cannot have a negative inside the square root. So set the insides greater-than-or-equal-to zero, and solve. This will be the domain:

$$-2x + \geq 30$$

$$-2x \geq -3$$

$$2x \leq 3$$

$$x \leq 3/2 = 1.5$$

Then the domain is "all $x \leq 3/2$".

The range requires a graph. Be sure to be careful when graphing radicals:

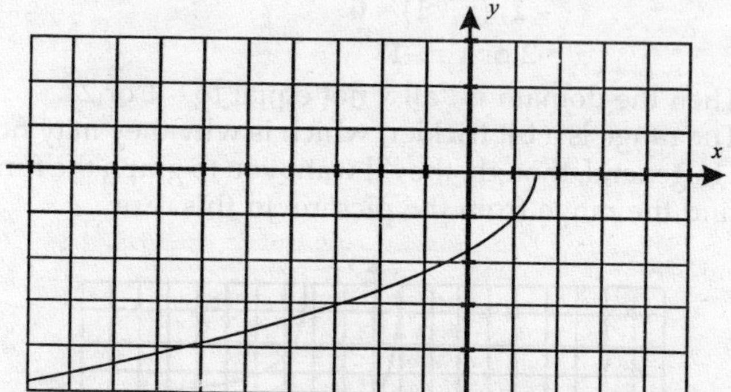

The graph starts at $y = 0$ and goes down from there. While the graph goes down very slowly, you know that, eventually, you can go as low as you like (by picking an x that is sufficiently big). Also, from your experience with graphing, you know that the graph will never start coming back up. Then the range is "$y \leq 0$".

Example: Determine the domain and range of the given function:

$$y = -x^4 + 4$$

This is just a garden-variety polynomial. There are no

denominators (so no division-by-zero problems) and no radicals (so no square-root-of-a-negative problems). There are no problems with a polynomial. There are no values that you can't plug in for x. When you have a polynomial, the answer is always that the domain is "all x".

The range will vary from polynomial to polynomial, and they probably won't even ask, but, if they do, look at the picture:

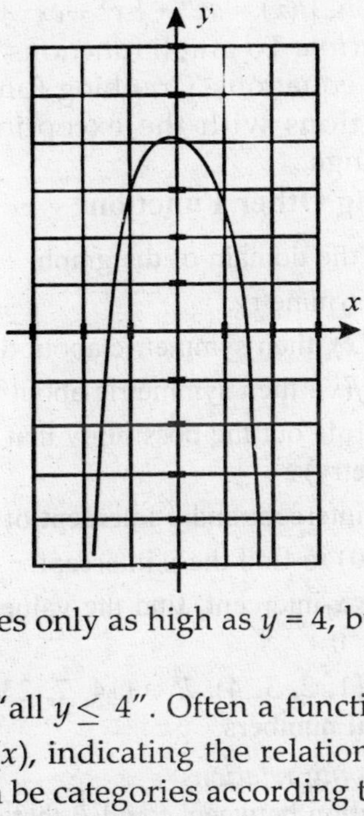

The graph goes only as high as $y = 4$, but it will go as low as you like. Then:

The range is "all $y \leq 4$". Often a function is denoted as $y = f(x)$ or simply $f(x)$, indicating the relation $\{(x, f(x)\}$.

Functions can be categories according to the degree of the functions, which are listed below:

Linear Function: A function of degree one is called linear function.

For example, $f(x) = 3x + 2$, the function is in the slope-intercept form ($y = mx + b$) m(slope) = 3, y-intercept = 2

Quadratic Function: A function of degree two is called quadratic function. [$f(x) = ax^2 + bx + c$]

The same as graphing quadratic equations. For example, graph: $f(x) = 2x^2 + 8x + 9$

$y = 2x^2 + 8x + 9$ replace $f(x)$ with y
$y - 9 = 2x + 8x$ complete the square
$y - 9 + 4 = 2\ (x^2 + 4x + 4)$
$y - 1 = 2(x + 2)^2$ parabola with vertex at (–2, 1)
$y = 2(x + 2)^2 + 1$ solve for y and replace y with $f(x)$
$f(x) = 2(x + 2)^2 + 1$

Cubic Function: A function of degree three is called cubic function. $[f(x) = ax^3 + bx^2 + cx + d]$

Biquadratic Function: A function of degree four is called biquadratic function. $[f(x) = ax^4 + bx^3 + cx^2 + dx + e]$

Graph of Function: To graph functions, just change y into $f(x)$ from graphing equations. Graphing functions is the same as graphing equations with the exception of naming the domain and the range.

Rules for Graphing Other Function:

1. Determine the domain of the graph.
2. Check for symmetry.
   if $f(-x) = f(x)$, then symmetric about the y-axis.
   if $f(-x) = -f(x)$, then symmetric about the origin.
   (functions rule out the possibility that the graph has an x-axis symmetry)
3. Find the x-intercept and y-intercept of the graph
   Evaluate $f(0)$ to find the y-intercept.
   To find the x-intercept, find the value or values that will make $f(x) = 0$

Examples: Let $A = \{1, 2, 3, 4)$, $B = (14, 7, 234)\}$, $C = \{a, b, c\}$, and R = real numbers.

Define the following relations:

1. r is the relation between A and B that associates the pairs $1 \sim 234, 2 \sim 7, 3 \sim 14, 4 \sim 234, 2 \sim 234$
2. f is the relation between A and C that relates the pairs $\{(1, c), (2, b), (3, a), (r, b)\}$
3. g is the relation between A and C consisting of the associations $\{(1, a), (2, a), (3, a)\}$
4. h is the relation between R and itself consisting of pairs $\{(x, \sin(x))\}$

Solution:

1. The relation r is not a function, because the element 2 from the set A is associated with two elements from B
2. The relation f is a function, because every element from A has exactly one relation from the set C.
3. The relation g is not a function, because the element $\{4\}$ from the domain A has no element associated with it.
4. The relation h is a function with domain R, because every element $\{x\}$ in R has, exactly one element $\{\sin(x)\}$ associated with it.

The outcomes of a function (i.e. the elements from the range associated to elements in the domain) do not only depend on the rule of the function (such as x is associated with $\sin(x)$) but also on the domain of the function. Therefore, we need to specify those outcomes that are possible for a given rule and a given domain:

Definition: Image and Preimage

Let A and B be two sets and f a function from A to B. Then the image of f is defined as

1. Image $(f) = \{b \in B$: there is an $a \in A$ with $f(a) = b\}$.

   Let A and B be two sets and f a function from A To B. If C is a subset of the range B then the preimage, or inverse image, of C under the function f is the set defined as
2. f^{-1} (C) = $\{x \in A: f(x) \in$ C$\}$

   Function can be classified into three groups: those for which every element in the image has one preimage, those for which the range is the same as the image, and those which have both of these properties. Accordingly, we make the following definitions:

   As an example, consider the following functions:

Example: Let $f(x) = 0$ if x is rational and $f(x) = 1$ if x is irrational. This function is called Dirichlet's Function. The range for f is R

1. Find the image of the domain of the Dirichlet Function when:

   (*a*) The domain of f is Q

(*b*) The domain of *f* is *R*
(*c*) The domain of *f* is [0, 1] (the closed interval between 0 and 1]

2. What is the preimage of *R*? What is the preimage of [–1/2, 1/2]?
3. What is the image of [0, 2] and the preimage of [1, 4]?
4. Find the image and the preimage of [–2, 2].

TYPES OF MAPPING

Mappings can be classified into a number of different types.

Injective

A function *f* from *A* to *B* is called one to one (or one-one) in whenever *f*(*a*) = *f*(*b*) then a = *b*. Such functions are also called injections.

Thus, the mapping $a : X \rightarrow Y$ is called injective or one-one if for each $x \in X$ there exists at most $y \in Y$ one such that $a(x) = y$. (Note that no such *x* may exist). Another way of saying that *a* is injective is $a(x)$ implies *y* where *x*. A graphical representation of an injective mapping is show in figure 1.

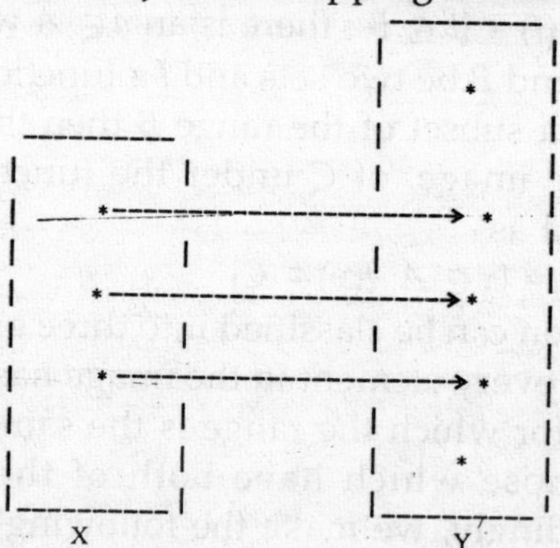

Fig. An Injective Mapping $a : X \rightarrow Y$ (not Surjective)

Surjective

A function *f* from *A* To *B* is called onto if for all *b* in *B* there is an *a* in *A* such that *f*(*a*) = *b*. Such functions are also called surjections.

Thus, the mapping $a : X \rightarrow Y$ is called surjective or onto if for each $x \in X$, there exists at least one $y \in Y$ such that $a(x) = y$. This type of mapping is illustrated in fig.

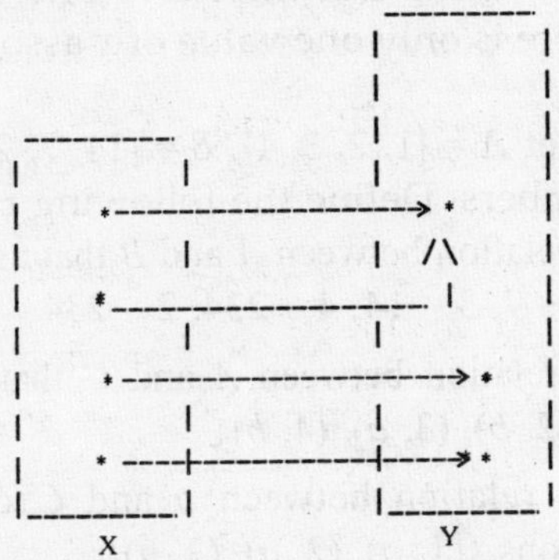

Fig. A Surjective Mapping $a : X \rightarrow Y$ (not Injective)

Bijective

A function f from A to B is called a bijection if it is one to one and onto, i.e. bijections are functions that are injective and surjective.

Thus, the mapping which is both injective and surjective is called bijective or one-to-one and such a mapping is illustrated in Figure. For such mapping, there is a one-to-one correspondence between the members of X and Y.

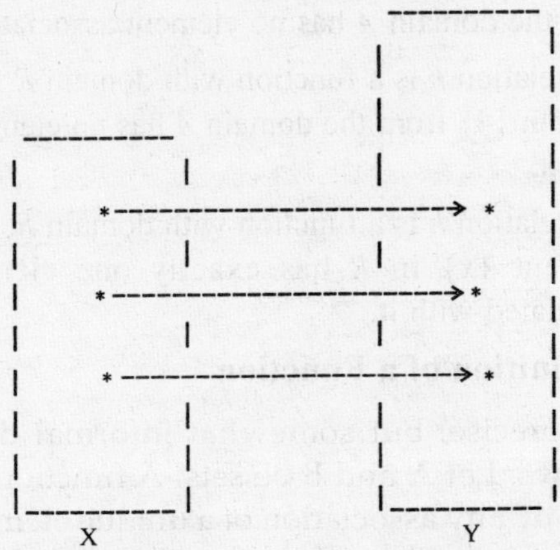

Fig. A bijective mapping $a : X \rightarrow Y$ (both injective and surjective)

Horizontal Line Test:

The horizontal line test implies a one to one function, meaning that there is only one value of x associated with each value of $f(x)$.

Examples: Let $A = \{1, 2, 3, 4\}$, $B = \{14, 7, 234\}$, $C = \{a, b, c\}$, and R = real numbers. Define the following relations:

1. r is the relation between A and B that associates the pairs $1 \sim 234 \sim 7$, $3 \sim 14$, $4 \sim 234$, $2 \sim 234$
2. f is the relation between A and C that relates the pairs $\{(1, c), (2, b), (3, a), (4, b)\}$
3. g is the relation between a and C consisting of the associations $\{(1, a), (2, a), (3, a)\}$
4. h is the relation between R and itself consisting of pairs $\{x, \sin(x)\}$

Which of those relations are functions?

Solution:

1. The relation r is not a function, because the element 2 from the set A is associated with two elements from B.
2. The relation f is a function, because every element from A has exactly one relation from the set C.
3. The relation g is not a function, because the element $\{4\}$ from the domain A has no element associated with it.
4. The relation h is a function with domain R, because every element $\{4\}$ from the domain A has no element associated with it.
5. The relation h is a function with domain R, because every element $\{x\}$ in R has exactly one element $\{\sin(x)\}$ associated with it.

Abstract Definition of a Function

A fairly precise, but somewhat informal definiton of a function follows. Let A and B be sets, A function from A to B is determined by any association of a unique element of B with each element of A. The set A is called the domain of the function; the set B is called the codomain.

Functions Versus Maps

A map is usually a (total) function as defined here, especially for authors using function as synonym for partial functions with a domain (of definition) not necessarily containing all points of the source or "departure set". (For them, $R \rightarrow R$; $x \rightarrow 1/x$ is a function with domain $R\backslash\{0\}$.)

Mathematical Definition of a Function

The informal idea of a function as a rule has been used since ancient times and is still used as the definition of a function in informal contexts such as introductory calculus textbooks. A typical example of this informal definition, is that a function is a rule that assigns to each element of a set D a single element of set C. This informal definition is sufficient for many purposes, but it relies on the undefined concept of a "rule". In the late 1900s, the question of what constitutes a valid rule defining a function came to the forefront. The consensus of modern mathematicians is that the word "rule" should be interpreted in the most general sense possible: as an arbitrary binary relation.

Thus it is common in advanced mathematics for an example to formally define a function f from a set D to a set C to be a set Gf of ordered pairs (x, y) in the Cartesian product $D \times C$. It is required that for each x in D there is at most one pair (x, y) in the set Gf. The "rule" of the function is: given x in D, if there is a pair (x, y) in Gf then $f(x) = y$, and otherwise $f(x)$ is not defined. The set of x for which f is defined is called the domain of f; if the domain of f is all of D then f is called total and the notation $f : D \rightarrow C$ is used. The set C is called the codomain of the function; this must be specified because it is not determined by Gf. The set Gf is called the graph of the function.

In most areas of mathematics, the word function is used to mean total function, although non-total functions (called partial functions) are important in functional analysis, mathematical logic, and category theory.

FUNCTIONS IN SET THEORY

Variations of this formal definition are sometimes more

convenient for specific disciplines. In some contexts of category theory, even if a function *f* from a set *D* is not defined for every element of *D*, the set *D* is still called the domain of *f*. *I* set theory, it is common to identify the function *f* with its graph *Gf*; this identification removes the need to specify either *D* or *C* in the formal definition.

Real Life Examples of Mappings

In order to capture the various concepts associated with mappings and to summaries, it will be helpful to look briefly at two examples of mappings.

Consider firstly a class of 25 students who sit an examination where marks are awarded in the range) to 100. If we denote the set of student names by *S* and the set of possible marks by *M* (so that *M* will be the set $\{m \mid m \in N, 0 \leq m)$, then the "examinations-results-list" constitutes a total mapping since every student has an examination mark and that mark is unique. We can write this mapping as

examination-results- $S \rightarrow M$

If every student whose name appears in *S* sat the examination, the mapping is total. Note that not every possible examination mark will appear on the results list because there are 101 possible marks but only 25 students. Hence for each mark, it is not the case there is at least one student name, who has attained that mark and so the mapping is not surjective. If it transpired that each of the 25 students achieved a different mark in the examination, then the mapping would be injective (since for each m, there is at most one student with that mark).

Suppose now that there are 200 students taking the examination and that each of the 101 possible marks has been scored. In this situation, for each mark, there is at least one student who has achieved that mark and so now the mapping is sujective. The mapping is not injective however since at least one mark will have been scored by more two or more students. In the (unlikely) event of 101 students taking the examination and each scoring a different mark, the resulting mapping would be bijective (both injective and surjective) and we would have a one-to-one correspondence between the set of student names and the set of examination marks. As a second example, consider a simple memory manager in an operating system. The heap contains memory available to applications programs and the space in the heap zone is subdivided into a number of contiguous blocks, say. Each block in the zone is either allocated to a user or is free. The storage manager is responsible for recording which blocks are currently owned by which user. If we denote the set of user-names by U and the set of blocks By H (so that H will be the set $\{m \mid m \in N, 0 \leq m \leq h_N\}$, then the storage manager constitutes a mapping from the set of block numbers to the set of users which we can express as

Storage-manager: Note that this constitutes a partial mapping in the situation where one or more blocks is not owned by any user. Also, since, in general, some users may not own any blocks while other users may own several, then the mapping will be neither surjective nor injective.

Chapter 4

Recurrence Relations

OVERVIEW

In mathematics, a recurrence relation is an equation that defines a sequence recursively: each term of the sequence is defined as a function of the preceding terms.

A difference equation is a specific type of recurrence relation.

An example of a recurrence relation is the logistic map:

$$x_{n+1} = rx_n (1 - x_n).$$

Some simply defined recurrence relations can have very complex (chaotic) behaviours and are sometimes studied by physicists and mathematicians in a field of mathematics known as nonlinear analysis.

Solving a recurrence relation means obtaining a closed-form solution: a non-recursive function of n.

LINEAR HOMOGENEOUS RECURRENCE RELATIONS

The term *linear* means that each term of the sequence is defined as a linear function of the preceding terms. The *order* of a linear recurrence relation is the number of preceding terms required by the definition—so the relation $a_n = a_{n-2}$ is of order two, because there must be at least two preceding terms (whether they are both used or not). The general form of a linear recurrence relation of order d is as follows:

$$a_n = c_1 a_{n-1} + c_2 a_{n-2} + ... + c_d a_{n-d} + c$$

where c and c_i (for all i) are allowed to depend on n, but a_i (for all) is not. If is independent of (for all), then the recurrence relation is said to have constant coefficients. Additionally, if then the recurrence relation is homogeneous; such a sequence is also called a linear recursive sequence or LRS.

The linear recurrence, together with *seed values* (initial conditions) for, determines the sequence uniquely.

FIBONACCI NUMBERS

The Fibonacci numbers are defined using the linear recurrence relation

$$F_n = F_{n-1} + F_{n-2}$$

with seed values:

$$F_1 = 1$$
$$F_2 = 1.$$

The sequence of Fibonacci numbers begins:

1, 1, 2, 3, 5, 8, 13, 21, 34, 55, 89...

It can be solved via the matrix solution described below, yielding the closed form expression.

Solving via Linear Algebra

Given an LRS, one can write down its companion matrix, then put it in Jordan normal form (which is diagonal if the eigenvalues are distinct). Expressing the seed in terms of the eigenbasis, say

$$\begin{bmatrix} a_0 \\ \mathrm{M} \\ a_{d-1} \end{bmatrix} = b_1 v_1 + \mathrm{K} + b_d v_d$$

yields

$$\begin{bmatrix} a_n \\ \mathrm{M} \\ a_{n+(d-1)} \end{bmatrix} = C_n \begin{bmatrix} a_0 \\ \mathrm{M} \\ a_{d-1} \end{bmatrix}$$

$$= C^n (b_1 v_1 + ... + b_d v_d) = \lambda_1^n b_1 v_1 + ... + \lambda_d^n b_d v_d$$

which is a closed form expression (expand on the first coordinate to obtain a closed form expression for a_n).

If the companion matrix is not diagonalizable, then the resulting expression is more complicated, but conceptually the same.

Solving Generally

Solutions to recurrence relations are found by systematic means, often by using generating functions (formal power series) or by noticing the fact that r^n is a solution for particular values of r. Consider, for example, a recurrence relation of the form

$$a_n = Aa_{n-1} + Ba_{n-2}.$$

Suppose that it has a solution of the form $a_n = r^n$. Substituting this guess in the recurrence relation, we find:

$$r^n = Ar^{n-1} + Br^{n-2}.$$

Dividing through by r^{n-2}, we get

$$r^2 = Ar + B,$$

$$r^2 = Ar - B = 0$$

This is known as the *characteristic equation* of the recurrence relation. Solve for r to obtain the two roots ë$_1$, ë$_2$, and if these roots are distinct, we have the solution

$$a_n = C\lambda_1^n + D\lambda_2^n$$

while if they are identical (when $A^2 + 4B = 0$), we have

$$a_n = C\lambda^n + D\lambda^n$$

where constants C and D can be found from the "side conditions" that are often given as $a_0 = a$, $a_1 = b$.

Different solutions are obtained depending on the nature of the roots of the characteristic equation.

GENERAL LINEAR HOMOGENEOUS RECURRENCE RELATIONS

Many linear homogeneous recurrence relations may be solved by means of the hypergeometric series. Special cases of these lead to recurrence relations for the orthogonal polynomials, and many special functions. For example, the solution to

$$J_{n+1} = \frac{2n}{z} J_n - J_{n-1}$$

is given by the $J_n = J_n(z)$, the Bessel function, while

$$(b-n)M_{n-1} + (2n-b-z)M_n - nM_{n+1} = 0$$

is solved by $M_n = M(n,b;z)$ the confluent hypergeometric series.

SOLVING NON-HOMOGENEOUS RECURRENCE RELATIONS

If the recurrence is inhomogeneous, a particular solution can be found by the method of undetermined coefficients and the solution is the sum of the solution of the homogeneous and the particular solutions. Another method to solve an inhomogeneous recurrence is the method of *symbolic differentiation*. For example, consider the following recurrence:

$$a_{n+1} = a_n + 1$$

This is an inhomogeneous recurrence. If we substitute $n \leftarrow n + 1$, we obtain the recurrence

$$a_{n+2} = a_{n+1} + 1$$

Subtracting the original recurrence from this equation yields

$$a_{n+1} - a_{n+1} = a_{n+1} - a_n$$

or equivalently

$$a_{n+2} = 2a_{n+1} - a_n$$

This is a homogeneous recurrence which can be solved by the methods explained above. In general, if a linear recurrence has the form

$$a_{n+k} = \lambda_{k-1}a_{n+k-1} + \lambda_{k-2}a_{n+k-2} + \ldots + \lambda_1 a_{n+1} + \lambda_0 a_n + p(n)$$

where $\lambda_0, \lambda_1, \ldots, \lambda_{k-1}$ are constant coefficients and $p(n)$ is the inhomogeneity, then if $p(n)$ is a polynomial with degree r, then this inhomogeneous recurrence can be reduced to a homogeneous recurrence by applying the method of symbolic differentiation r times.

Example: Let's solve the following inhomogeneous recurrence relation using the method of undetermined coefficients:

$$a_{n+1} = 2a_n + 3^n + 5_n$$

The general solution for the analogous homogeneous relation

$$b_{n+1} = 2b_n \text{ is}$$

$$b_n = c_1 2^n$$

The inhomogeneous part ($3^n + 5n$) leads us to guess the particular solution

$$a_n = c_2 a + c_3 n + c_4$$

(the guess for a^x is a^x, and the guess for a degree-n polynomial is a degree-n polynomial)

Substituting back into the recurrence relation, we get:

$$c_2 3a^{n+1} + c_3(n + 1) + c_4$$

$$= 2(c_2 3^n + c_3 n + c_4) + 3^n + 5n$$

Looking only at the coefficients of 3^n:

$$3c_2 = 2c_2 + 1$$

$$c_2 = 1$$

Looking only at the coefficients of n:

$$c_3 = 2c_3 + 5$$

$$c_3 = -5$$

Looking only at the coefficients of 1 (constants):

$$c_3 + c_4 = 2c_4$$

$$c_4 = c_3 = -5$$

So the general solution is

$$a_n = c_1 2^n + 3^n - 5n - 5$$

A technique is described for using Newton's identities to compute the coefficients of a multi-step recurrence formulae for simulating the solutions of ordinary linear differential equations of high order. This approach gives the exact root-matched recurrence formula without having to determine the complex roots of the characteristic polynomials.

Let x(t) and y(t) denote continuous functions of time related according to the Nth order differential equation

$$\sum_{k-0}^{N} a_k \frac{d^k y}{dt} = \sum_{k=0}^{N} b_k \frac{d^k x}{dt^k} \qquad a_N, a_0, b_N {}^1 0$$

where the coefficients a_k and b_k are constants. Simple linear relationships of this form are widely used in numerical simulations of heat transfer, chemical reactions, fluid flows, mechanical dynamics, Markov processes, etc. (Relationships of this form are also used in signal filtering and control-system compensation, but in those applications the frequency response is usually emphasized over the exact time-domain response.) A variety of methods have been used to numerically generate the solutions y(t) of such equations for arbitrary forcing functions x(t). In the digital environment equation is often simulated by means of a multi-step linear recurrence relation of the form

$$h\,Y_n = g\,X_n$$

where Xn and Yn are the column vectors

$$Y_n = \begin{bmatrix} y\big((n-N)\Delta t\big) \\ \vdots \\ y\big((n-N)\Delta t\big) \\ y(n\Delta t) \end{bmatrix} \quad X_n = \begin{bmatrix} y\big((n-N)\Delta t\big) \\ \vdots \\ y\big((n-N)\Delta t\big) \\ x(n\Delta t) \end{bmatrix}$$

for some constant Δt (called the *step-size* of the simulation), and g and h are constant row vectors

$$h = [h_0\ h_1 \dots h_N]\ g = [g_0\ g_1 \dots g_N]$$

Given the coefficients ak and bk, our objective is to compute, as quickly and efficiently as possible, the components of h and g.

It is well known that if the recurrence is required to match the exact homogeneous response for y(t) when x(t) is constantly zero, then the components of h must be proportional to the elementary symmetric functions (with alternating signs) of the quantities $e^{\alpha_i \Delta t}$, i = 1,2,..,N, where αi are the roots of the characteristic polynomial.

The usual procedure for determining this "root-matched" recurrence is to solve the characteristic polynomial for the roots ai, and then compute the elementary symmetric functions of these roots. For high order relationships the most troublesome aspect of this approach is the determination of all the complex roots of the characteristic polynomials, especially when there are multiple roots.

The most common way of simplifying the calculations is to use a substitution method, e.g., bilinear substitution, but such methods do not reproduce the exact homogeneous response. The purpose of this note is to describe how, using

Newton's Identities, we can precede directly from the coefficients to the exact root-matched recurrence coefficients without solving the characteristic polynomials.

ROOT MATCHED RECURRENCE

For any integer k, let w(k) denote the sum of the kth powers of the roots $a_1, a_2, .., a_N$. Clearly, w(0) = N. Using Newton's Identities [8,9] the values of w(k) for k = 1,2,... are given by

$a_N\, w(1) + 1a_{N-1} = 0$

$a_N\, w(2) + a_{N-1}\, w(1) + 2a_{N-2} = 0$

$a_N\, w(3) + a_{N-1}\, w(2) + a_{N-2}\, w(1) + 3a_{N-3} = 0$

for k < N, and the recurrence

$a_N\, w(k) + a_{N-1}\, w(k-1) + ... + a_0\, w(k-N) = 0$

for all k ≥ N. Now let E(k) denote the sum of the kth powers of the values of $e^{\alpha_i \Delta t}$, i=1,2,..,N. Expanding each of the exponential functions as a Taylor's series and combining terms by powers of Δt gives the following expression for E(k):

$$E(k) = \sum_{j-0}^{\infty} \frac{(k\Delta t)^j}{j!} w(j)$$

Letting σ_k denote (-1)k times the kth elementary symmetric function of the quantities $e^{\alpha_i \Delta t}$, we can now compute σ_0 through σ_N using the identities

$E(1) + 1\sigma_1 = 0$

$E(2) + \sigma_1\, E(1) + 2\sigma_2 = 0$

$E(3) + \sigma_1\, E(2) + \sigma_2\, E(1) + 3\sigma_3 = 0$

The precision of these computations, including the convergence, can be checked by verifying the equality

$$\sigma_N = (-1)^N\, e^{-a_{N-1}\Delta t / a_N}$$

The general recurrence for the homogeneous equation in y(t) can now be expressed as:

$$S_A Y_n = 0$$

where SA is the row vector

$$S_A = [\sigma_N\ \sigma_{N-1} \cdots \sigma_0]$$

Figure presents an implementation of the preceding

algorithm in BASIC. The inputs to this programme are the order N, the step-size Dt, and the coefficients $a_0, a_1, .., a_N$. The programme computes the values of $\sigma_0, \sigma_1, .., \sigma_N$. The parameter "sumto" determines the number of terms to be evaluated in the summation.

By entering the coefficients bk in place of ak into the preceding algorithm we can determine the recurrence for the homogeneous equation in x(t)

$$S_B X_n = 0$$

We can then combine the two sides of the recurrence by applying the appropriate scale factors, which are easily deduced from steady-state conditions. The result is

$$\frac{a_0}{\sum s_A} S_A Y_n = \frac{b_0}{\sum s_B} S_B X_n$$

where $\Sigma\, S_A$ and $\Sigma\, S_B$ denote the sums of the components of SA and SB respectively.

EQUILATERAL RECURRENCES

If $b_N = 0$ then the characteristic polynomial of the right side has fewer than N roots. In this case the recurrence would involve fewer "past values" of x(t) than of y(t). Note that the coefficients of x_n were uniquely determined by the requirement that x(t) conform to the true homogeneous behaviour when y(t) is constantly zero.

However, in practice x(t) is an arbitrary forcing function so we are not really concerned with reproducing the homogeneous "response" of x(t). Instead, we are trying to reconstruct the function x(t) from the discrete samples. Thus it is often considered more appropriate to make use of as many past values of x(t) as is computationally convenient. This typically leads to recurrences that use the same number of past values of x(t) as of y(t), i.e., *equilateral* recurrences. The usual approach in signal processing is to augment the roots of the right-hand characteristic polynomial with fictitious roots of the form ±p i, and then proceed as before. The effect is to add roots equalling -1 to the characteristic polynomial of the recurrence, which essentially mimics bi-linear substitution.

The rationale for the "method of fictitious roots" just described is based on the frequency response characteristics. In the time domain it is often more appropriate to convert into an equilateral equation by a simple change of variables. If we define the variable z(t) such that y=λ (x+z) for some constant λ, then (1) can be written

$$\sum_{k-0}^{N} a_k \frac{d^k z}{dt^k} = \sum_{k-0}^{N} \left(\frac{b_k}{\lambda} - a_k \right) \frac{d^k x}{dt^k}$$

For any choice of λ not equal to 0, (b_N/a_N), or (b_0/a_0), this equation is equilateral, so we can apply the method to give the recurrence

$$h\, Z_n = g\lambda\, X_n$$

where gλ is based on the coefficients of the right hand side of (5). Reverting back to the y(t) variable gives

$$h\, Y_n = \lambda\, (g\lambda + h)\, X_n$$

For large values of λ the vector λ (gλ + h) is quite insensitive to changes in λ, and it converges in the limit as $|\lambda| \to \infty$. In practice, any λ greater than about 10 times the largest b_i will give very nearly the limiting vector. This "change-of-variables method" enables us to determine equilateral recurrences for non-equilateral differential equations without resorting to fictitious roots.

VALUE METHOD

For comparison we briefly describe a method that explicitly treats x(t) as an *independent* variable and y(t) as a *dependent* variable. From this point of view it can be argued that the *expected value* of x(t) based on the N+1 most recent samples is given by the Nth degree polynomial that passes through those samples. Although the method is a straight-forward application of polynomial curve fitting, no explicit formulation of the general Nth-order recurrence formula appears in the literature, so a brief description is given below.

As in our previous recurrence formulas, we require that the homogeneous response of y(t) be matched exactly, so we use the algorithm shown in Figure to determine SA. Combining this with the particular solution based on the Nth order polynomial curve fit of x(t) leads directly to the total recurrence

$$S_A Y_n = S_A T A^{-1} B T^{-1} X_n$$

where A, B, and T are square matrices such that the elements in the jth columns of the kth rows (where j,k range from 0 to N) are given by

$$A_{kj} = \begin{cases} \frac{j!}{k!} a_{j-k} & \text{if } j \ge k \\ 0 & \text{if } j \ge k \end{cases} \qquad B_{kj}$$

$$B_{kj} = \begin{cases} \frac{j!}{k!} b_{j-k} & \text{if } j \ge k \\ 0 & \text{if } j < k \end{cases}$$

$$T_{kj} = \begin{cases} (k\Delta t)^j & \text{if } k + j > 0 \\ 0 & \text{if } k + j = 0 \end{cases}$$

This method gives, arguably, the best possible reconstruction of arbitrary forcing functions x(t), although the matrix inversions make it somewhat laborious when applied to high order equations.

ROOT-MATCHED RECURRENCES

Here is a method for determining the "root-matched" recurrence corresponding to a given ordinary differential equation without having to explicitly solve for the roots of the characteristic polynomial. More generally, this method enables us to find the exact root-matched recurrence corresponding to any continuous linear system without having to solve for the eigen values of the system. The following description covers the just simple homogeneous case. Suppose you have an ordinary linear differential equation with constant coefficients:

$$a5\, y''''' + a4\, y'''' + a3\, y''' + a2\, y'' + a1\, y' + a0\, y = 0$$

where the (') denotes derivatives with respect to t. The response of y(t) can be reproduced discretely by a recurrence relation of the form

$$c5\, y(5dt) + c4\, y(4dt) + c3\, y(3dt) + c2\, y(2dt) + c1\, y(dt) + c0\, y(0) = 0$$

where "dt" is the time increment of the simulation. The question is, given the coefficients ai of the differential equation, how can you compute the coefficients ci of the corresponding recurrence relation.

The coefficients ci must be proportional to the elementary symmetric functions (with alternating signs) of the quantities $e^{(ri\ dt)}$, where ri (i=1,2,..,5) are the roots of the characteristic polynomial. Thus, a straightforward method of computing the ci is to solve the characteristic polynomial for the roots ri, and then compute the elementary symmetric functions of these roots. However, determining all the roots of a high order polynomial can be computationally tricky, particularly for embedded code, where the use of iterative processes can cause problems.

Fortunately, there is a way to compute the ci without having to solve for the characteristic roots of the equation. The technique makes use of "Newton's Identities" and the Taylor's series expansion of the exponential function.

Assume our basic ODE is of order N. For any integer k, let w(k) denote the sum of the kth powers of the roots r1,r2,...,rN. Clearly, w(0)=N. Using Newton's Identities, the values of w(k) for k=1,2,... are given by

w(1)a(N) + 1a(N–1) = 0

w(2)a(N) + w(1)a(N–1) + 2a(N–2) = 0

w(3)a(N) + w(2)a(N–1) + w(1)a(N–2) + 3a(N–3) = 0 etc

for k=N. Now let E(k) denote the sum of the kth powers of the values $e^{(ri\ dt)}$, i=1,2,..,N. Expanding each exponential function as a Taylor series and combining terms by powers of dt gives the following expression for E(k):

E(k) = SUM from j=0 to inf of {((k dt)^j) w(j)/ j!}

for k=1,2,..,N. The coefficients of the recurrence relation can then be computed using the identities

c(N) = 1

E(1) + 1 c(N–1) = 0

E(2) + E(1) c(N – 1) + 2 c(N–2) = 0

E(3) + E(2) c(N–1) + E(1) c(N–2) + 3 c(N–3) = 0

This gives the exact root-matched recurrence coefficients

without ever having to determine the roots. It also has the advantage that no special provisions are needed to cover the case of repeated roots.

GOLDEN PENTAGON

The plane can be tiled by a geometric progression of equilateral triangles with the characteristic equation $x^3 - x - 1$ as shown below. In view of the popularity of the "golden rectangle" based on the geometric spiral of squares, it's surprising the analagous construction for equilateral triangles is rarely mentioned. Another picture, this one with the triangles numbered insequence, is shown below.

Notice that if h(k) denotes the height of the kth trianglein the spiral, then each of the following recurrences issatisfied

$h(k) = h(k\text{-}1) + h(k–5)$

$h(k) = h(k\text{-}2) + h(k–3)$

This corresponds to the fact that the sequence has two possiblecharacteristic equations

$x^5 - x^4 - 1 = 0$

$x^3 - x - 1 = 0$

This isn't surprising, because the second polynomial is adivisor of the first, i.e.,

$(x^5 - x^4 - 1) = (x^3 - x - 1)(x^2 - x - 1)$

Each triangle in the sequence is larger than the

preceedingtriangle by a factor w, defined as the real root of thecubic factor $x^3 - x - 1$. This value can be expressed as

$$w = \left\langle \frac{1}{2} + \left\langle \frac{1}{2^2} - \frac{1}{3^3} \right\rangle^{1/2} \right\rangle^{1/2} + \left\langle \frac{1}{2} - \left\langle \frac{1}{2^2} - \frac{1}{3^3} \right\rangle^{1.2} \right\rangle^{1/3}$$

The numerical value of this "other golden proportion" is

$$w = 1.324717957244746...$$

Chapter 5

Combinatories

HISTORY

The appearance of the Fibonacci number five in prosody. There is one way to arrange one beat, two for two, three for three, and five for four beats.

The earliest books about combinatories are from India· A Jainist text, the Bhagabati Sutra, had the first mention of a Combinatories problem; it asked how many ways one could take six tastes one, two, or three tastes at a time. The Bhagabati Sutra was written around 300 BC, and thus was the first book to mention the choice function· The next ideas of Combinatories came from Pingala, who was interested in prosody. Specifically, he wanted to know how many ways a six syllable meter could be made from short and long notes. He wrote this problem in the Chanda sutra (also Chandahsutra) in the second century BC· In addition, he also found the number of meters that had n long notes and k short notes, which is equivalent to finding the binomial coefficients.

The ideas of the Bhagabati were generalized by the Indian mathematician Mahariva in 850 AD, and Pingala's work on prosody was expanded by Bhaskara and Hemacandra in 1100 AD. Bhaskara was the first known person to find the generalized choice function, although Brahmagupta may have known earlier· Hemacandra asked how many meters existed of a certain length if a long note was considered to be twice as long as a short note, which is equivalent to finding the Fibonacci numbers·

While India was the first nation to publish results on Combinatories, there were discoveries by other nations on similar topics. The earliest known connection to Combinatories comes from the Rhind papyrus, for the implementation of a geometric series. The next milestone is held by the I Ching. The book is about what different hexagrams mean, and to do this they needed to know how many possible hexagrams there were. Since each hexagram is a permutation with repetitions of six lines, where each line can be one of two states, solid or dashed, Combinatories yields the result that there are $2^6 = 64$ hexagrams. A monk also may have counted the number of configurations to a game similar to Go around 700 AD· Although China had relatively few advancements in enumerative Combinatories, they solved a combinatorial design problem, the magic square, around 100 AD.

ENUMERATIVE COMBINATORIES

Counting the number of ways that certain patterns can be formed is the central problem of enumerative Combinatories. Two examples of this type of problem are counting combinations and counting permutations. More generally, given an infinite collection of finite sets $\{S_i\}$ indexed by the natural numbers, enumerative Combinatories seeks to describe a *counting function* which counts the number of objects in S_n for each n. Although counting the number of elements in a set is a rather broad mathematical problem, many of the problems that arise in applications have a relatively simple combinatorial description.

The simplest such functions are *closed formulas,* which can be expressed as a composition of elementary functions such as factorials, powers, and so on. For instance, as shown below, the number of different possible orderings of a deck of n cards is $f(n) = n!$. Often, no closed form is initially available. In these cases, we frequently first derive a recurrence relation, then solve the recurrence to arrive at the desired closed form. Finally, $f(n)$ may be expressed by a formal power series, called its *generating function,* which is most commonly either the ordinary generating function

$$\sum_{n\geq 1} f(n)x^n$$

or the exponential generating function

$$\sum_{n\geq 1} f(n)\frac{x^n}{n!}.$$

Often, a complicated closed formula yields little insight into the behaviour of the counting function as the number of counted objects grows. In these cases, a simple asymptotic approximation may be preferable. A function $g(n)$ is an asymptotic approximation to $f(n)$ if $f(n)/g(n) \to 1$ as $n \to$ infinity. In this case, we write $f(n) \sim g(n)$ Once determined, the generating function may allow one to extract all the information given by the previous approaches. In addition, the various natural operations on generating functions such as addition, multiplication, differentiation, etc., have a combinatorial significance; this allows one to extend results from one combinatorial problem in order to solve others.

Permutations with Repetitions

When the order matters, and an object can be chosen more than once, the number of permutations is n^r where n is the number of objects from which you can choose and r is the number to be chosen. For example, if you have the letters A, B, C, and D and you wish to discover the number of ways to arrange them in three letter patterns (trigrams)

1. Order matters (e.g., A-B is different from B-A, both are included as possibilities)
2. An object can be chosen more than once

you find that there are 4^3 or 64 ways. This is because for the first slot you can choose any of the four values, for the second slot you can choose any of the four, and for the final slot you can choose any of the four letters. Multiplying them together gives the total.

FIBONACCI NUMBERS

Let $f(n)$ be the number of distinct subsets of the set $S(n) = \{1, 2, 3,..,n\}$ that do not contain two consecutive integers. When $n = 4$, we have the sets {}, {1}, {2}, {3}, {4}, {1,3}, {1,4}, {2,4}, so $f(4) = 8$.

We count the desired subsets of $S(n)$ by separately counting those subsets that contain element n and those that do not. If a subset contains n, then it does not contain element n “ 1. So there are exactly $f(n$ “ 2) of the desired subsets that contain element n. The number of subsets that do not contain n is simply $f(n$ “ 1). Adding these numbers together, we get the recurrence relation:

$$f(n) = f(n-1) + f(n-2),$$

where $f(1) = 2$ and $f(2) = 3$.

As early as 1202, Leonardo Fibonacci studied these numbers. They are now called Fibonacci numbers; in particular, $f(n)$ is known as the n + 2nd Fibonacci number. Although the recurrence relation allows us to compute every Fibonacci number, the computation is inefficient. However, by using standard techniques to solve recurrence relations, we can reach the closed form solution:

$$f(n) = \frac{\phi^{n+2} - (1-\phi)^{n+2}}{\sqrt{5}}$$

where $\phi = (1 = \sqrt{5})/2$, the golden ratio.

In the above example, an asymptotic approximation to $f(n)$ is:

$$f(n) \sim \frac{\phi^{n+2}}{\sqrt{5}}$$

as n becomes large.

Here the questions are of the following type: what is the probability of a certain graph property for a random graph (within a certain class) E.g. what is the average number of triangles in a random graph?

Probabilistic methods are also used to determine the existence of combinatorial objects with certain prescribed properties (for which explicit examples might be difficult to find), simply by observing that the probability of randomly selecting an object with those properties is greater than 0.

PARTITIONS INTO DISTINCT PARTS

For any positive integers n and k, let $p_k(n)$ denote the number of ways in which the integer n can be expressed as a sum of exactly k distinct positive integers, without regard to order. For example, the integer $n = 12$ can be expressed as a sum of three distinct positive integers in the following seven ways:

1+2+9 = 1+3+8 = 1+4+7 = 1+5+6 = 2+3+7 = 2+3+7 = 2+4+6 =3+4+5

Therefore we have $p_3(12) = 7$. Obviously $p_1(n) = 1$ for all n, so the generating function for $p_1(n)$ is just the geometric series

$$\frac{x}{1-x} 1x^1 + 1x^2 + 1x^3 + \ldots$$

The values of $p_2(n)$ for $n = 1, 2, 3, \ldots$ are easily seen to be 0, 0, 1, 1, 2, 2, 3, 3, 4, 4, ... and so on. In other words, $p_2(1+2j) = p_2(2+2j) = j$. Factoring x^3 and $1+x$ from the power series with these coefficients, we get

$$1x^3 + 1x^4 + 2x^5 + 2x^6 \ldots = x^3(1+x)(1+2x^2+3x^4+4x^6 \ldots)$$

The infinite series on the right is just the derivative of the previous geometric series, except with x^2 in place of x, so we can differentiate the closed form of the geometric series and insert into this expression to give the generating function for $p_2(n)$

$$\frac{x^3(1+x)}{(1-x^2)^2} = \frac{x^3}{(1-x)(1-x^2)}$$

The first several values of $p_3(n)$ are listed (by rows) below.

0	0	0	0	0	1
1	2	3	4	5	7
8	10	12	14	16	19
21	24	27	30	33	37
40	44	48	52	56	61

Just as every second term of the $p_2(n)$ series gives an

arithmetic sequence, we expect that the values of $p_3(n)$ should give sequences of degree 2, and this is indeed the case, as can be seen by taking every sixth term. The values in the columns are given by second-degree polynomials, i.e., we have

$$p_3\,(1+6j) = 3j^2 - 2j$$

$$p_3\,(2+6j) = 3j^2 - 1j$$

$$p_3\,(3+6j) = 3j^2$$

$$p_3\,(4+6j) = 3j^2 + 1j$$

$$p_3\,(5+6j) = 3j^2 + 2j$$

$$p_3\,(6+6j) = 3j^2 + 3j + 1$$

The generating functions for these individual sub-sequences are

$$x^2\,\frac{1+5x^6}{(1-x^6)^3} = 0x^1 + 1x^7 + 8x^{13} + 21x^{19} + \dots$$

$$x^8\,\frac{2+4x^6}{(1-x^6)^3} = 0x^2 + 2x^8 + 10x^{14} + 24x^{20} + \dots$$

$$x^9\,\frac{3+3x^6}{(1-x^6)^3} = 0x^3 + 3x^9 + 12x^{15} + 27x^{21} + \dots$$

$$x^{10}\,\frac{4+2x^6}{(1-x^6)^3} = 0x^4 + 4x^{10} + 14x^{16} + 30x^{22} + \dots$$

$$x^{11}\,\frac{5+1x^6}{(1-x^6)^3} = 0x^5 + 5x^{12} + 16x^{17} + 33x^{23} + \dots$$

$$x^6\,\frac{1+4x^6+x^{12}}{(1-x^6)^3} = 1x^1 + 7x^{12} + 19x^{18} + 37x^{24} + \dots$$

The sum of these sub-sequences gives the generating function for the entire $p_3(n)$ sequence. The numerator of the sum is a (shifted) palindromic polynomial

$$x^6 + x^7 + 2x^8 + 2x^9 + 4x^{10} + 5x^{11} + 4x^{12} + 5x^{13} + 4x^{14} + 3x^{15}$$
$$+2x^{16} + x^{17} + x^{18} = x^6\,(x+1)^2\,(x^2+x+1)^2\,(x^2-x+1)^3$$

Dividing this by the denominator and simplifying, we get

the complete generating function for the number of partitions of n into 3 distinct parts

$$\frac{x^6}{(1-x)(1-x^2)(1-x^3)} = 1x^6 + 1x^7 + 2x^8 + 3x^9 + 4x^{10}$$
$$+5x^{11} + 7x^{12} + \ldots$$

Notice that if x^6 is replaced with 1 in the numerator the resulting expression would be the generating function for the number of partitions of n into positive integers (not necessarily distinct) no greater than 3. We can see this immediately because each of the factors in the denominator contributes a geometric series in x^k where k is 1, 2, and 3 respectively, and hence the coefficient of x^n for n greater than 0 is the number of ways in which n is the sum of those three elements. Therefore, letting $P_j(n)$ denote the partitions of n into positive integers no greater than j (and defining $P_3(0) = 1$), we have shown that $P_3(n) = p_3(n + 6)$ for all n. This correspondence can be represented pictorially as shown below for the case $n = 5$.

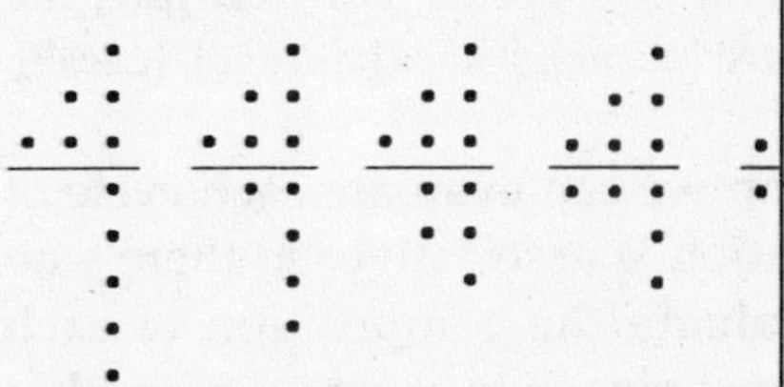

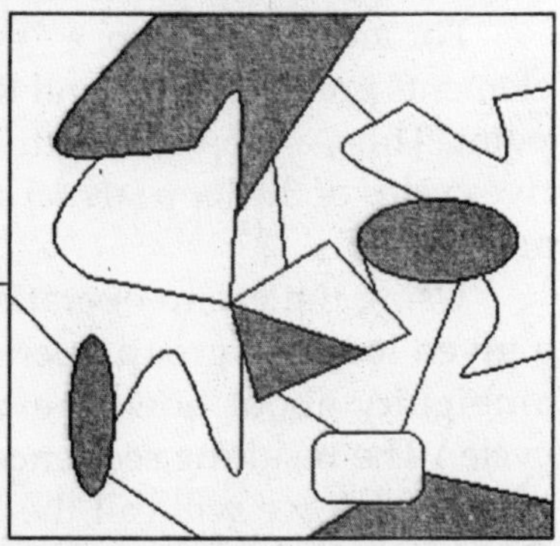

Each partition of 11 into distinct positive integers must have a smallest part of at least 1, a second smallest part of at least 2, and a third smallest part of at least 3, so we can begin with this triad of 6 dots, and then to each column we add zero or more dots, with the requirement that the number of added dots can never decrease as we go from the left-most to the right-most column. The total number of additional dots to be added is 5, and with the stated restriction these rows can be interpreted as the partitions of 5 into positive integers no greater than 3. By the same reasoning it follows that, for any positive integer k, we have $P_k(n) = p_k(n + k(k+1)/2)$. In other words, the number of partitions of n into positive integers no greater than k is equal to the number of partitions of $n + k(k + 1)/2$ into k distinct positive integers. Thus the generating function for the latter is

$$\frac{x^{k(k+1)2}}{(1-x)(1-x^2)\text{L}\ (1-x^k)}$$

PERMUTATION

Suppose a, b, c, d, e are elements of a group. We'll call a circular arrangement of such elements a "loop". For example, we might choose to arrange those elements in alphabetical order, giving the loop {*abcde*}, where the last element is understood to wrap around to the first. Note that {*abcde*} and {*eabcd*} are both valid representations of the same loop, since they are just rotations of each other.

For any given loop, a "cycle" consists of the sequence of elements going once around the loop from any specific starting point. Thus, a loop of length N has N cycles. For example, the five cycles of the loop {*abcde*} are [*abcde*], [*eabcd*], [*deabc*], [*cdeab*], and [*bcdea*].

Using the group operation we can evaluate each cycle of a given loop. (A group operation is associative, so there's no ambiguity about how to evaluate the composition of each cycle.) The resulting sequence of elements constitutes another loop, which we will call the "successor" of the original loop. Of course, if the group operation is commutative, each cycle of a given loop will evaluate to the same element, so a successor loop is not very interesting. However, if the group operation is NOT commutative, then successor loops will typically be non-trivial.

Even if the group operation is not commutative, it's clear that the elements of a successor loop must all belong to a single conjugacy class. To see why, suppose the cycle [*abcde*] evaluates to the group element x. The next cycle is [*eabcd*], which can be written as $e\ x\ e^{-1}$, so it is conjugate to x. If we call this new element y, then the next element is [deabc], which can be written as d $y\ d^{-1}$, so it too is in the conjugacy class of x and y. And so on.

It's interesting to see how the successor operation affects loops consisting of elements of the symmetric group S_n. I'll represent the elements of S_n by the permutations of n objects, and accordingly I'll refer to these loops as "permutation loops". Also, I'll focus on the effects of the successor operation on *complete* permutation loops, consisting of some arrangement of all n! elements in a loop of length n!.

The elements of S_2 are just the two permutations represented by I=[12] and A=[21]. There's only one complete permutation loop of order 2, namely, {IA}, which is the same loop as {AI}. The first successor of this loop is {AA}, and the next successor is {II}. All subsequent successors are simply {II}.

For a slightly less trivial example, consider the symmetric group S_3, whose sixe elements are

I = 123	C = 231
A = 132	D = 312
B = 213	E = 321

One possible loop of these elements is {IDEACB}. The six cycles of this loop are evaluated as

IDEACB = A
DEACBI = A
EACBID = B
ACBIDE = A
CBIDEA = A
BIDEAC = E

Therefore, the successor of {IDEACB} is {AABAAE}. Continuing in this way we can generate the sequence of successors as shown below:

{IDEACB}, {AABAAE}, {DCDCDC}, {IIIIII}, {IIIIII},...

FOUR COLOUR THEOREM

How many different colors are sufficient to colour the countries on a map in such a way that no two adjacent countries have the same colour? The figure below shows a typical arrangement of colored regions.

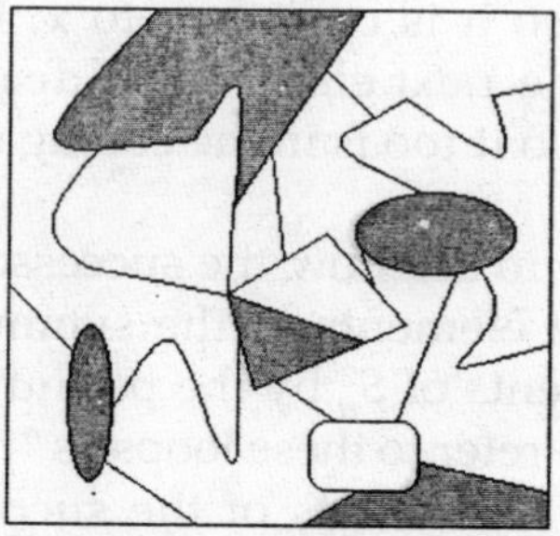

Notice that we define adjacent regions as those that share a common boundary of non-zero length. Regions which meet at a single point are not considered to be "adjacent".

After examining a wide variety of different planar graphs, one discovers the apparent fact that every graph, regardless of size or complexity, can be colored with just four distinct colors. This "four-colour conjecture" was first noted by August Ferdinand Mobius in 1840.

In 1852 a young man named Francis Guthrie wrote about the problem in a letter to his brother Frederick, then a student at University College in London. Neither of the brothers was able to prove the conjecture, so Frederick asked one of his professors, Augustus DeMorgan.

DeMorgan too was unable to prove the conjecture, and after recognizing the difficulty of the problem, he wrote to Sir William Rowan Hamilton to ask for help.

It might seem that this problem would be irresistible to Hamilton, since anything having to do with the number "four" immediately suggests a connection with his beloved quaternions. Also, Hamilton made contributions to graph theory (such as the idea of a Hamiltonian circuit, i.e., a path along the edges of a graph that visits each vertex exactly once), a subject that was developed largely through efforts to prove the four colour conjecture. Nevertheless, Hamilton immediately wrote back that he did not believe he would solve DeMorgan's "quaternion of colors" any time soon.

The coloring of geographical maps is essentially a topological problem, in the sense that it depends only on the connectivities between the countries, not on their specific shapes, sizes, or positions. We can just as well represent each country by a single point (vertex), and the adjacency between two bordering countries can be represented by a line (edge) connecting those two points. It's understood that connecting lines cannot cross each other. A drawing of this kind is called a planar graph. A simple map (with just five "countries") and the corresponding graph are shown below.

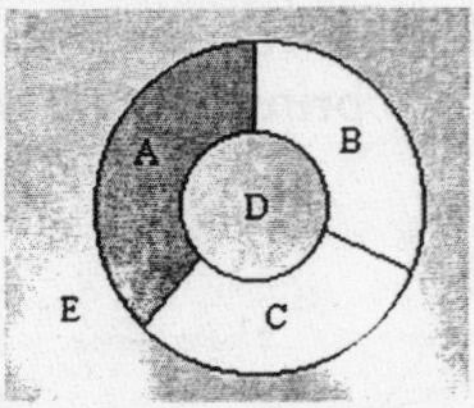

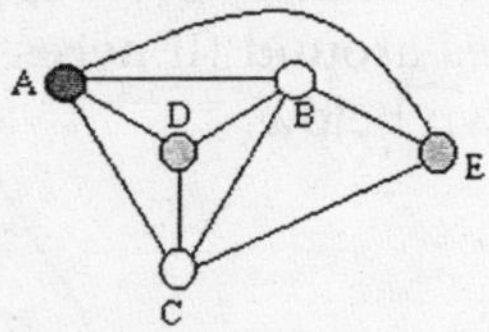

A graph is said to be n-colorable if it's possible to assign one of *n* colors to each vertex in such a way that no two connected vertices have the same colour. Obviously the above graph is not 3-colorable, but it is 4-colorable. The Four Colour Theorem asserts that every planar graph–and therefore every "map" on the plane or sphere–no matter how large or complex, is 4-colorable.

Despite the seeming simplicity of this proposition, it was only proven in 1976, and then only with the aid of computers.

Notice that the above graph is "complete" in the sense that no more connections can be added (without crossing lines).

The edges of a complete graph partition the graph plane into three-sided regions, i.e., every region (including the infinite exterior) is bounded by three edges of the graph. Every graph can be constructed by first constructing a complete graph and then deleting some connections (edges). Clearly the deletion of connections cannot cause an n-colorable graph to require any additional colors, so in order to prove the Four Colour Theorem it would be sufficient to consider only complete graphs.

Although DeMorgan was unable to prove the four-colour

conjecture, he did observe that no more than four regions on the plane can all be in mutual contact with each other. The graph of a set of three mutually adjoining regions is simply a topological triangle, and if we add a fourth region, it is represented by a fourth vertex in the graph, which must be located either inside or outside the triangle formed by the graph of the original three vertices.

In either case, we must then connect this fourth vertex with each of the three original vertices so that all four of the regions are mutually adjoining. Having done this, we can slide the vertices around (if necessary) to bring the graph into the form shown below.

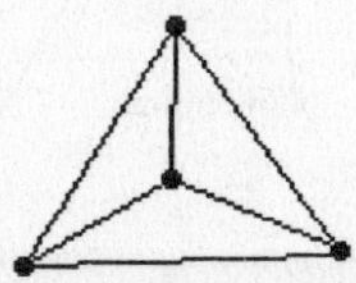

This is the unique graph of four mutually adjoining plane regions (V = 4, E = 6), and also of the tetrahedron. It clearly divides the graph plane into four isolated regions, one region exterior to the big triangle, and the three interior regions. A fifth vertex added to this graph will be able to have edges that reach only three of the four existing vertices, because each of the four regions is completely bounded by three edges that block access to one of the existing vertices.

Hence there does not exist a plane graph of five mutually connected vertices, so there does not exist a set of five mutually adjoining regions on the plane.

Of course, the non-existence of a graph with five mutually adjoining vertices does not automatically imply the non-existence of a graph requiring five distinct colors. For example, there exist graphs in which the largest subset of mutually adjoining vertices is two, and yet for which three colors are required.

This is the case with any simple loop with an odd number of vertices, as illustrated in the left-hand figure below. Similarly there are graphs containing no subset of four mutually adjoining vertices that nevertheless require four colors, such as the graph shown on the right.

Therefore, we must consider the possibility that five colors might be required for some graph, even though it's not possible for a graph to contain any set of five mutually connected vertices.

On the other hand, it could be argued that the two example shown above can each be reduced, clarifying the causal links between different parts of the graphs. For example, in the left figure we have an alternating sequence of red, green, red, green, and if these were the only two available colors, it's clear that the chain must continue to alternate, no matter how long–or how short. Thus we could just as well replace any chain of the form rgrg...rg with the simple pair rg, in which case the left-hand figure above consists of three mutually connected vertices. Likewise if we replace the circumferential chain rgrg around the central yellow vertex in the right hand figure with just the pair rg, we get a set of four mutually connected vertices.

In view of this, it's tempting to think that DeMorgan's observation somehow, perhaps indirectly, does ultimately account for the truth of the four-colour theorem. Indeed if we construct a graph by adding vertices, one at a time, and make the maximum number of connections at each stage, we will always find the graph plane divided into "triangular" regions, each of which has access to only three vertices. Thus it follows that four colors suffice for any graph that can be constructed in this way. For example, consider the two graphs shown below.

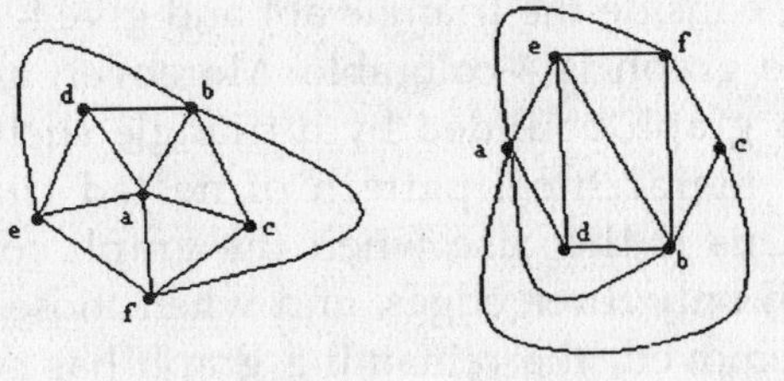

These two graphs each have V = 6 vertices and E = 12 edges, and in fact the two graphs are topologically identical. The plane regions represented by the vertices "*a*" and "*b*" each have five adjoining neighbors, whereas the vertices "e" and "f" each have four, and the vertices "*c*" and "*d*" each have three. These are complete graphs and, as noted above, a complete graph divides the entire graph plane into "triangular" regions, i.e., regions bounded by three edges connecting three vertices. Nevertheless, depending on how we add the points, it is possible that when a vertex is added to a graph it has more than three neighbors, so we cannot say automatically that four colors would suffice. For example, if vertex "a" is the last to be added, it would have five pre-existing neighbors, and if four colors have already been used in those five vertices, the vertex "a" would require a fifth colour.

However, we need not add vertex "*a*" last. Another topologically equivalent way of drawing the above graph is shown below

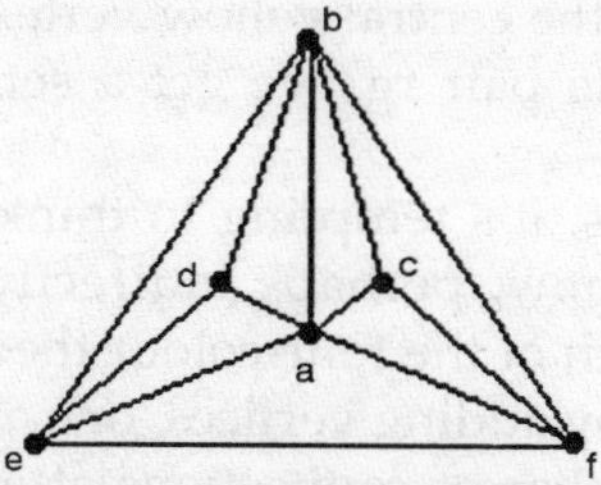

This shows that we could first assign three distinct colors to the vertices e,b,f, and then place the vertex "*a*" in this triangle, connect it to each of the three surrounding vertices, and give it a fourth colour. Then we can place vertex d inside the triangle abe and give it the same colour as *f*. Then we can place vertex c inside the triangle abf and give it the colour of e. Hence the graph is 4-colorable. Moreover, any graph, or portion of a graph bounded by a triangle such as ebf, and having this hierarchical pattern of nested triangles, is 4-colorable. This is the case when the graph contains some vertices with only three edges, and when those vertices and edges are removed, the remaining graph has some vertices with only three edges, which can be removed, and so on, until finally all that remains is a single triangle.

Unfortunately (for the prospect of a simple proof), not every graph is of this hierarchical form. For example, consider the complete graph shown below.

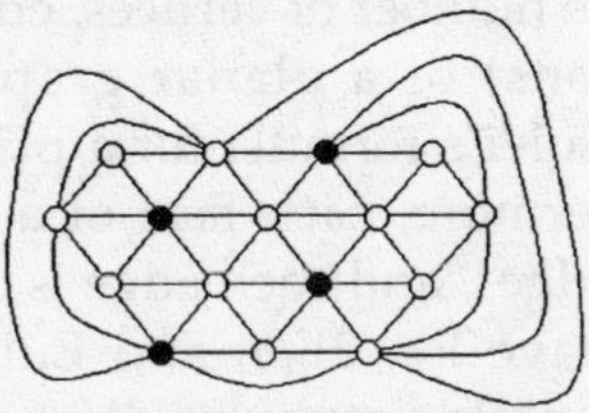

This graph has V = 16 vertices and E = 42 edges. It is not hierarchical, because after removing the two vertices having only three edges each, the remaining graph has four or more edges attached to each vertex. (We can also infer that this graph is not hierarchical from the fact that no three mutually connected vertices are directly connected to a fourth vertex.) Nevertheless, this graph is 4-colorable, as shown in the figure.

The above graph contains a "flat" patch consisting of a uniform hexagonal grid. An *infinite* hexagonal grid is obviously 4-colorable by means of the alternating layered pattern shown below.

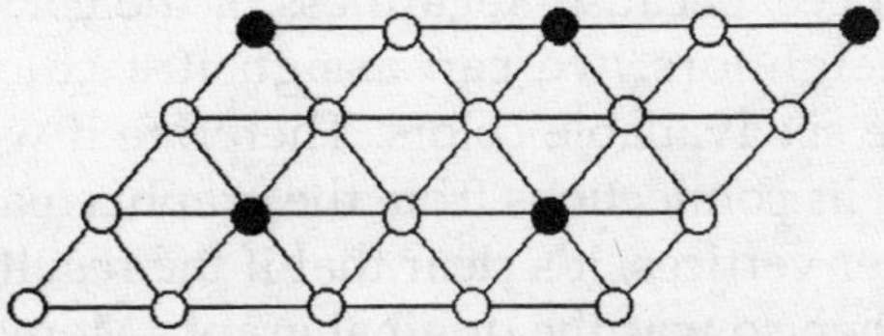

In a sense, two extreme types of graph structures are the purely hierarchical graphs and the perfectly "flat" graphs, both of which are easily seen to be 4-colorable.

Notice that we need not consider graphs containing any configurations of four mutually connected vertices, because, as discussed previously, the edges of such a configuration necessarily split the graph plane into four separate regions, each of which has access to only three of the four vertices.

Hence if we can 4-colour the interior of such a triangle, we can 4-colour the tetrahedral "frame" as well. If we could reduce all the causal links in a graph to their bare minimums, it might be possible to reduce every n-colorable graph to a graph with just n mutually connected vertices, and hence n could never exceed four. However, the reduction of graphs with more than two colors is not a trivial task.

If we denote the number of vertices, edges, and faces (i.e., the bounded regions) of a planar graph by V, E, and F respectively, then Euler's formula for a plane (or a sphere) is V–E + F = 2. Furthermore, each face of a complete graph is bounded by three edges, and each edge is on the boundary of two faces, so we have F = 2E/3, and Euler's formula for a complete planar graph is simply E = 3V–6. Now, each edge is connected to two vertices, so the total number of attachments (in a complete graph) is 2E = 6V–12, and hence the average number of attachments per vertex is 6–12/V. For any incomplete graph, the total number of attachments is less. Consequently the average number of attachments per vertex for any graph (with a finite number of vertices) is less than 6, which implies that at least one vertex has only five or fewer attachments.

If we have six available colors, a vertex with only five neighbors obviously imposes no constraint on the coloring of the other vertices, because, regardless of the colors of its five (or fewer) neighbors, we can assign it a colour without exceeding the six available colors. Therefore if we delete this vertex and all its connections from the graph, creating a graph with one fewer vertices, it's clear that if the resulting graph is 6-colorable, then so was the original graph. Moreover, Euler's formula assures us that this reduced graph also contains at least one vertex with five or fewer neighbors, so we can apply this procedure repeatedly, reducing the graph eventually to one with just 6 vertices, which is obviously 6-colorable. Hence the original graph is 6-colorable.

So, we've seen that Euler's formula immediately implies that no graph can require more than six colors. Furthermore, with just a little more work, we can also show that no graph can require more than *five* colors. (Ultimately we will see that no graph can require more than four colors, but it's worthwhile to begin with the proof of the 5-colorability of every planar graph.) Obviously every graph with five or fewer vertices is 5-colorable, so if there exists a finite graph that requires more than five colors, it must have more than five vertices. Let the positive integer V_6 denote the smallest number of vertices on which there exists a graph that requires six (or more) colors. Conversely, every graph with fewer than V_6 vertices is 5-colorable.

Now, assuming the existence of a graph that requires more than five colors, we can consider one that has exactly V_6 vertices. By Euler's formula, this graph must contain at least one vertex with five or fewer connections. However, it cannot contain any vertex with just four (or fewer) connections, because if it did, we could delete such a vertex and leave a graph with just V_6–1 vertices, which is 5-colorable by definition. Re-inserting the deleted vertex would clearly have no effect on the 5-colorability of the graph, because the vertex has only four (or fewer) neighbors, so the original graph must be 5-colorable, contradicting our assumption. Therefore, a graph with V_6 vertices that requires more than five colors cannot contain any vertex with just four or fewer neighbors.

Since Euler's formula implies that the graph contains at least one vertex with five or fewer connections, the only remaining possibility is that the graph contains a vertex with exactly five connections. However, this too leads to a contradiction. To show this, it's helpful to introduce the notion of a *k-cluster*, which is specified by a set of *k* distinct colors and one particular vertex that has one of those colors. The original vertex is included in the cluster, and, in addition, every vertex with one of the *k* specified colors that neighbors a vertex in the cluster is also in the cluster. By definition the only vertices outside a cluster that are directly connected to vertices inside the cluster have colors that are not in the specified set of *k* colors. Therefore, we can apply any permutation of the *k* colors to the vertices in a cluster without invalidating the coloration. In particular, a 2-cluster is a contiguous set of vertices, each with one of two specified colors, and we can transpose these two colors without upsetting the coloration of a graph. Now consider a graph containing a vertex with exactly five immediate neighbors, of five distinct colors, as illustrated below.

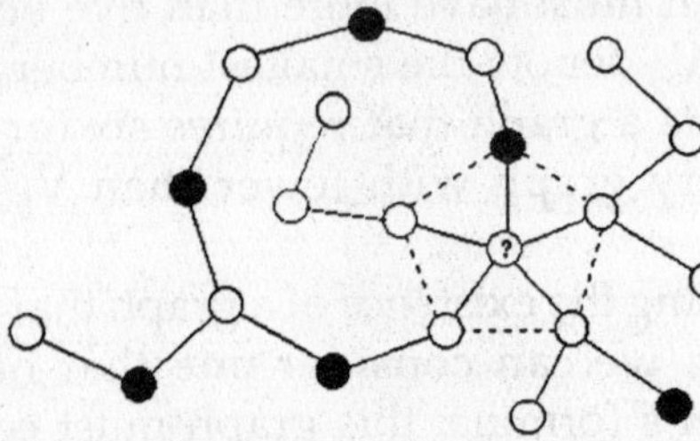

Since the uncolored vertex in the centre has neighbors of five distinct colors, it might seem that a sixth colour is required. However, notice that we can transpose the blue and green colors in the blue/green 2-cluster attached to the upper left vertex of the central pentagon, so that the upper left vertex is green instead of blue. Once we have done this, the uncolored vertex in the centre has neighbors of only four distinct colors. If we delete the central vertex, the overall graph has V_6–1 vertices, so it is 5-colorable, but re-inserting this vertex requires no sixth colour (once we have transposed the blue and green in the small 2-cluster as described), so the original graph is 5-colorable.

The only possible objection to the transposing of the colors in a cluster as a means of reducing the number of distinct colors around the perimeter of the pentagon is that we cannot necessarily change the colour of any vertex to the colour of one of the opposite vertices of the pentagon, because two opposite vertices of a pentagon might be in the same 2-cluster. This is the case, for instance, with the yellow and red vertices of the pentagon in the figure above. If we transpose the red and yellow colors in this 2-cluster, there would still be five distinct colors around the perimeter of the pentagon. However, if such a cluster exists, connecting two opposite vertices of the pentagon, we are guaranteed that at least one other pair of vertices are cut off from each other, i.e., they cannot be in the same 2-cluster. For example, the blue/green 2-cluster outlined in red in the above figure cannot be part of the blue/green 2-cluster attached to the upper right of the pentagon. In general, for any vertex connected to exactly five other vertices, we can always (without invalidating the coloration) change the colour of at least one of the five neighbors so that there are only four distinct colors. It follows, using the reduction argument described in the preceding paragraph, that no graph can require more than five colors.

This still leaves open the question of whether five colors are ever actually required, or whether four colors will always suffice. If there exist planar graphs requiring five distinct colors, there must be a positive integer V_5 that is the smallest number of vertices on which such a graph can be constructed. Let's call a graph with V_5 vertices that requires five colors a minimal 5-colour graph. By the reduction argument described previously, such a graph obviously cannot contain any vertex with three or fewer connections. Moreover, by considering 2-clusters again, we can prove that such a graph cannot contain any vertex with just four connections. To see why, consider the portion of a graph illustrated below.

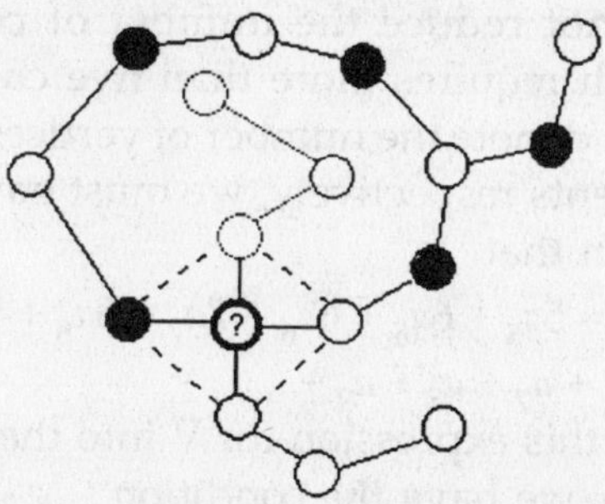

The uncolored vertex in the centre has just four neighbors, which we've colored red, yellow, green, and blue. This might seem to require a fifth colour for the central vertex, but in fact by permuting the yellow and blue colors in the 2-cluster (outlined in red) above the central vertex we can transform this coloration so that the central uncolored vertex has neighbors of only three distinct colors (red, green, and blue).

Since the overall graph from which this configuration was taken is assumed to have exactly V_5 vertices, it follows that if we delete the central vertex, the resulting graph has only V_5–1 vertices, so it is 4-colorable, but we can clearly add the central vertex back without requiring a fifth colour, so the whole graph must be 4-colorable, contradicting our original assumption. This shows that it's always possible to modify a graph containing a vertex with only four connections in such a way that the vertex is connected to only three distinct colors. At most, only one of the two pairs of opposite vertices can be linked by a common 2-cluster, because if one pair is so linked, the other cannot be linked.

To summarize the argument up to this point, we've shown that no graph can require more than five colors, and we've also shown that if there exists a graph requiring five colors, there must exist a minimal 5-colour graph, and such a graph cannot contain any vertex with fewer than five connections. Also, by adding connections, we can "complete" any graph on V vertices so that all the faces are triangular and there are a total of exactly 6V-12 attachments.

This graph must still be a minimal 5-colour graph because we haven't increased the number of vertices, and adding connections cannot reduce the number of colors required, whereas no graph requires more than five colors. Therefore, letting a_5, a_6, a_7,... denote the number of vertices with precisely 5, 6, 7,... attachments respectively, we must have a minimal 5-colour graph such that

$$6V-12 = 5a_5 + 6a_6 + 6a_6 + 7a_7 + 8a_8 + 9a_9....$$

where $V = a_5 + a_6 + a_7 + a_8 + a_9 + ...$

Substituting this expression for V into the above formula and re-arranging, we have the condition

$$12 = a_5 - a_7 - 2a_8 - 3a_9 - ....$$

This places fairly severe constraints on any putative complete minimal 5-colour graph. For example, if we consider only such graphs with no more than six attachments at any single vertex, then this formula implies $a_5 = 12$. In other words, a complete minimal 5-graph with no more than six attachments per vertex must have exactly 12 vertices with five attachments each. This suggests that these 12 vertices are arranged globally in the pattern of an icosahedron, and the remaining vertices with six attachments per vertex are in a regular hexagonal pattern, filling in the "faces" of the icosahedron. (This pattern is the basis for "geodesic domes"). The fundamental graph of this type, with $a_6 = 0$, is shown below.

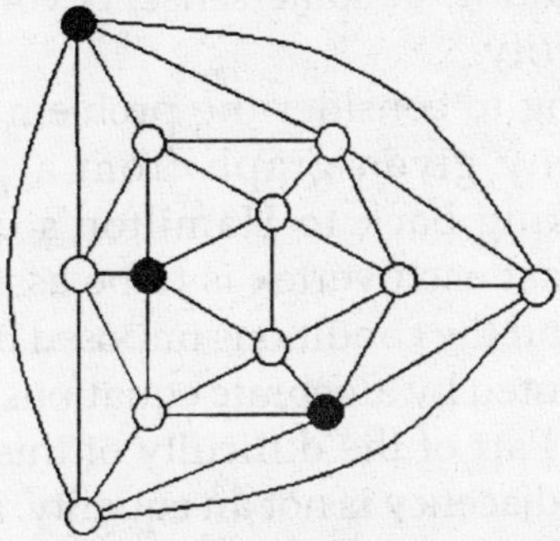

We can give explicit 4-colorings of every such pattern, so such graphs can be shown explicitly to require no more than four colors. Therefore, if a complete minimal 5-colour graph exists, it must contain at least one vertex with seven or more attachments.

It was not until 1976 that, with the help of modern computers, the 4-colour conjecture was finally proven to be true. The proof, developed by Appel and Haken based on ideas of several earlier people (Kempe, Heawood, Birkhoff, etc) consisted of finding a set of subgraphs, at least one of which must be contained in any normal planar graph (i.e., the set is unavoidable), and such that if a graph containing one of those sub-graphs is not 4-colorable, then neither is a graph with one fewer vertices. This leads immediately to a contradiction, because it implies that, if we hypothesize the existence of a graph that is not 4-colorable, we can always construct another graph, with fewer vertices, that is also not 4-colorable, eventually arriving at a graph with only four vertices, but this graph obviously *is* 4-colorable. Hence we have a contradiction, so we can conclude that the original hypothesis was false, i.e., there does *not* exist a graph that is not 4-colorable. The unavoidable set found by Appel and Haken consisted of nearly

1500 subgraphs, and many of these required considerable analysis to prove that they were "reducible". The development of this unavoidable set, and the proofs of reducibility for its members, was carried out largely on computers, so the proof is notable (and slightly controversial) as an early example of a mathematical proposition whose proof can only be carried out on a computer, and is, in some sense, beyond human mental capabilities to verify.

It's interesting to consider the problem of determining a 4-coloring for any given graph from a purely algebraic standpoint (harking back to Hamilton's quaternions). We might stipulate that each vertex is to be assigned one of four possible values, and the conditions imposed by the connections would be represented by algebraic equations, one equation for each connection. Part of the difficulty of this approach is that the condition of adjacency is not an equality, it is an *in*equality, i.e., we require that two vertices have unequal values (from among the four possible values). Also, there is necessarily a great deal of ambiguity in the solution, because any permutation of the colors leaves a solution intact. In addition, there is even more ambiguity, since in general there can be more than one distinct 4-coloration of a graph.

Given the four allowable (distinct) values a, b, c, d, we could algebraically impose the requirement for every vertex to have one of these values by requiring that the value u assigned to any vertex satisfies $f(u) = 0$, where the polynomial f is defined as

$$f(x) = (x-a)(x-b)(x-c)(x-d)$$

Then we could algebraically impose the required inequality on the values u,v assigned to two connected vertices by stipulating that $g(uv) = 0$, where the polynomial g is defined as

$$g(x) = (x-ab)(x-ac)(x-ad)(x-bc)(x-bd)(x-cd)$$

The equation g(uv) = 0 is satisfied if and only if u and v have distinct values from the set {a,b,c,d}. For example, the set {-2,-1,1,2} gives

$$f(x) = (x^2-1)(x^2-4) \qquad g(x) = (x+1)(x+4)(x^2-4)$$

To illustrate, consider again the octahedral graph (which is 3-colorable):

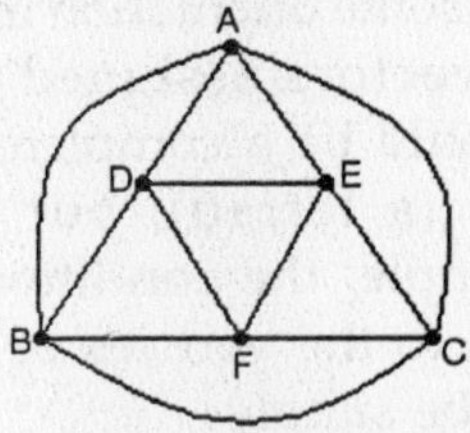

The algebraic conditions for the individual "colour" values assigned to the six variables A,B,C,D,E,F are

$$f(A) = f(B) = f(C) = f(D) = f(E) = f(F) = 0$$

and the algebraic conditions imposed by the twelve connections are

$$\begin{aligned} g(AB) &= g(AC) = g(AD) = g(AE) \\ &= g(BC) = g(BD) = g(BF) = g(CE) \\ &= g(FC) = g(DE) = g(DF) = g(EF) = 0 \end{aligned}$$

This amounts to 18 equations in the 6 variables. We can, however, without loss of generality stipulate the "colors" of three of the variables, say, A = 1, B = -1, C = 2, since these three must have mutually distinct values. This leaves us with the following twelve equations in the three unknowns

$$f(D) = f(E) = f(F) = 0$$

and

$$\begin{aligned} g(D) &= g(E) = g(-D) = g(-F) = g(2E) \\ &= g(2F) = g(DE) = g(DF) = g(EF) = 0 \end{aligned}$$

The three equations $f(D) = 0$, $g(D) = 0$, and $g(-D) = 0$ jointly have only two solutions, namely, D = 2 and D = -2. Therefore those three equations can be replaced with the single equation $D^2 - 4 = 0$. Likewise the three equations involving only E can be reduced to $E^2 + 3E + 2 = 0$, and the three equations involving only F can be reduced to $F^2 + F - 2 = 0$. Thus we have six equations in the three unknowns.

A different approach would be to assign a four-dimensional vector to each vertex, and then the connection between two given vertices would be represented by requiring the dot product of those two vectors to vanish, i.e., requiring that the two vectors be mutually orthogonal. The three vectors assigned to the vertices of any triangle would then be an orthogonal triad with some orientation in 4-dimensional space. Likewise the four vectors assigned to the vertices of a tetrahedral graph would be a complete set of four mutually orthogonal vectors (a tetrad), but still with arbitrary orientation. For example, the conditions on the six vectors A,B,C,D,E,F assigned to the vertices of the octahedral graph shown above would be simply

$$\begin{aligned} A \cdot B = A \cdot C = A \cdot D = A \cdot E \\ = B \cdot C = B \cdot D = B \cdot F = C \cdot E \\ = C \cdot F = D \cdot E = D \cdot F = E \cdot F = 0 \end{aligned}$$

In this context the four colour theorem tells us that a space of four dimensions is sufficient to enable us to assign one of the four basis vectors to each vertex of a planar graph in such a way that the vectors of every pair of adjacent vertices are orthogonal.

We can assume each vector is of unit length, so it has three independent components. Therefore, we have 18 independent components in all, constrained by 12 equations. Naturally the system is underspecified, because we can apply an arbitrary permutation to the vectors.

Since the edges of a complete graph partition the graph plane into three-sided regions whose vertices are three mutually connected points, we can uniquely assign the fourth colour to each of these regions.

COMBINATIONS AND INTERVAL GRAPHS

Let n continuous segments with freely defined starting and ending points be placed along the x axis, and consider the intersections between these segments. For any given arrangement there may be some regions where no segment is present, some where just one segment is present, some where exactly two segments overlap, some where exactly three segments overlap, and so on. Any such arrangement establishes a set of "meetings" between two or more segments. To illustrate, the figure below shows the placements of six segments, and identifies the meetings between them.

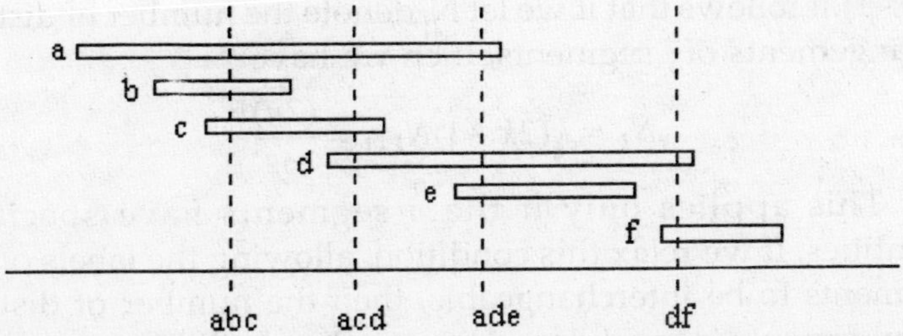

The meetings for this arrangement are abc, acd, ade, and df. In general, a meeting is a local maximum point on the plot of the number of segments present as a function of position. The plot for the above arrangement of six segments is shown in the figure below.

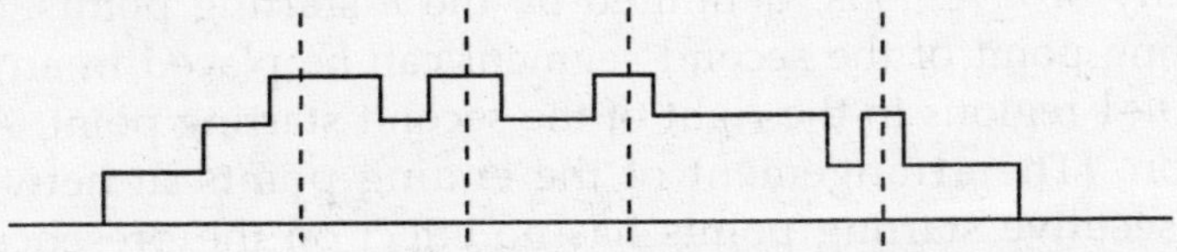

The number of different arrangements of the starting and stopping points for n segments (with distinct identities) is easy to determine. For the first segment there is obviously only one arrangement.

This segment divides the axis into three regions, namely, the region to the left of the starting point, the region between the starting and ending points (i.e., the region covered by the segment), and the region to the right of the ending point. Therefore, the starting point of a second segment can be placed in any one of those three regions, and the ending point can be placed in the same region or any other region to the right of the region containing the starting point.Thus there are six distinct ways of placing the second segment.

In general, as we prepare to add the kth segment, there will already be k–1 segments in place, and the 2(k–1) terminal points of these segments divide the axis into 2k–1 regions. The starting point for the kth segment can be placed in any of these, and then the ending point is placed in the same or more rightward region. Thus the number of possibilities for placing both these points is $(2k-1) + (2k-2) + ... + 2 + 1$, which equals

$k(2k–1)$.It follows that if we let N_j denote the number of distinct arrangements of j segments, then we have

$$N_k = k(2k-1)N_{k-1} = \frac{(2n)!}{2^n}$$

This applies only if the n segments have specified identities. If we relax this condition, allowing the labels of the segments to be interchangeable, then the number of distinct arrangements is much less. In general we will have n starting points, so it only remains to place the n ending points. The only restriction is that each ending point must follow its starting point. Thus the ending point for the first segment (i.e., the segment with the left-most starting point) can be placed in any of n regions, delimited by the n starting points. The ending point of the second segment can be placed in any of the $n–1$ regions to the right of the second starting point. And so on. (The arrangement of the ending points in between consecutive starting points has no effect on the presence of intersections.) Consequently the number of arrangements for indistinguishable segments is simply (n!).

However, this still distinguishes between the 1st segment, the 2nd segment, and so on. It is possible for two or more distinct arrangements to yield the same "meeting structure" up to permutation of the segments. For a trivial example, consider the two arrangements of three segments shown below.

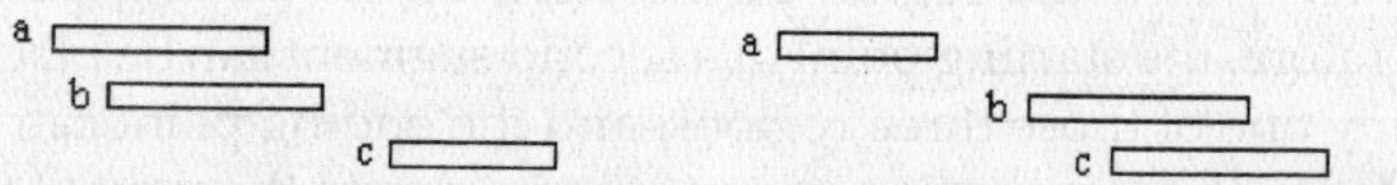

(The segment labels here merely indicate the order of the starting points, not implying any distinguishable identity for the segments.) In the arrangement shown on the left, the meeting structure is "ab", which signifies that the first and second segments overlap. In the arrangement on the right, the meeting structure is "*bc*", which signifies that the second and third segments overlap. If we don't care about the orderings of the start times for the segments in the meeting structure, then these two arrangements induce the same meeting structure. On this basis, what is the number of distinct meeting structures for n segments?

Let $M(n)$ denote the number of distinct meeting structures for n segments. Obviously for $n = 0$ and $n = 1$ segments, no meetings are possible, so there is only one meeting structure in each of these cases, namely the null structure, denoted by {Æ}. Thus we have M(0) = M(1) = 1. For $n = 2$ segments there are two possible meeting structures, corresponding to the fact that either these segments overlap or they do not. We will let {Æ, ab} denote these two possibilities, and we have M(2) = 2.

For $n = 3$ segments, there are 3! = 6 distinct arrangements, which we can express symbolically as

$$\varnothing \quad ab \quad bc \quad ab + bc \quad ab + ac \quad abc$$

However, as noted above, the arrangements *ab* and *bc* represent the same meeting structure. Likewise the arrangement *ab* + *bc* is the same structure as *ab* + *ac*. In both cases, one segment meets with each of the other two, but the other two don't overlap with each other. Therefore, the number of distinct meeting structures for $n = 3$ segments is just four, consisting of the set {Æ, *ab*, *ab* + *bc*, *abc*}, and so we have M(3) = 4. At this point we should make it clear that although our notation for meeting structures looks like Boolean logical notation, it is not. For example, the meeting "*abc*" can also be written in the form "*ab* + *ac* + *bc*", because these both signify a triple intersection between the segments a, b, and c. In this notation, conjoined characters signify a mutual meeting of those segments, and the "+" symbol represents logical "AND".

Thus both of these operations are forms of "intersection". It's important to note that if *a,b,c* represented arbitrary sets, then *ab* + *ac* + *bc* would *not* be equivalent to abc, because it would be possible to have those three pair-wise intersections with no three-way intersection. However, since our basic sets are continuous segments on the one-dimensional x axis, the three pairwise meetings of three segments imply that the first ending point occurs after the last starting point, so there is a region of three-way intersection. (It's interesting that the topology of continuous regions in one-dimensional space is induced by imposing the equality *ab* + *ac* + *bc* = *abc*, raising the question of whether other geometrical contexts can be specified in terms of some other Boolean equalities.) On the other hand, the "meeting" relation between segments is not transitive, because $ab + bc$ does not imply ac.

For $n = 4$ segments, we examine all $n! = 24$ distinct arrangements and find the following ten distinct meeting structures

(1)	$\varnothing$	(1)	$ab(c+d)$
(3)	ab	(3)	$a(bc+d)$
(4)	$a(b+c)$	(4)	$ab+bc+cd$
(1)	$ab+cd$	(1)	abc
(2)	$a(b+c+d)$	(2)	$abcd$

(The number to the left of each structure is the number of occurrences of that structure in the n! arrangements.) Hence we have M(4) = 10.

To determine the value of M(5), we could examine each of the 5! = 120 arrangements by hand, and check for the unique structures, but this would be very laborious. To automate the process, we can step through each of the n! placements of the n ending points, and for each arrangement we can easily determine the pairwise meetings. We can then count how many pairs contain the first segment, how many the second, and so on. We can then count how many segments appear in exactly one pair, how many in exactly two, and so on. This gives a set of numbers $d_1, d_2,...$ that characterize the meeting structure, independent of the orderings of the segments. If two structures are characterized by different values of d_j, then they are certainly distinct structures, so we can determine a lower bound on the value of M(n) by determining the number of unique characteristics $\{d_1, d_2,...\}$.

For the case $n = 5$, this criterion turns out to be nearly sufficient to distinguish all the structures. There are 26 distinct sets of $\{d_j\}$ indices, so we know M(5) is at least 26. However, it is possible for distinct meeting structures to have the same set of $\{d_j\}$ values. This occurs for the structures $ab + ac + bc + de$ and $ab + bc + cd + de$. In each case, two letters appear exactly once, and the other three letters appear twice each. However, they are obviously different structures, as seen by the fact that the first one reduces to abc+de whereas the second is irreducible. So, there are actually M(5) = 27 distinct meeting structures with $n = 5$. These are shown below.

$\varnothing$	$a(b+c+d)+de$	$a(ac+de)$
ab	$a(bc+d)$	$a(b(c+d)+e)$
$ab+cd$	$abc+d(a+e)$	$a(b(c+d)+de)$
$a(b+d)$	$a(b+c+d)+c(d+e)$	$a(bcd+e)$
$a(b+c)+de$	$ab(c+d)$	$ab(c+d+e)$
$ab+ac+cd$	$ab(c+d)+de$	$ab(cd+e)$
abc	$abcd$	$abc(d+e)$
$abc+de$	$a(b+c+d+e)$	$ab+bc+cd+de$
$a(b+c+d)$	$a(bc+d+e)$	$abcde$

To help distinguish structures that have the same $\{d_j\}$ sets, consider again the two structures ab+ac+bc+de and ab+bc+cd+de. If we replace each letter by the number of pairs in which it appears, these become 2·2 + 2·2 + 2·2 + 1·1 and 1·2 + 2·2 + 2·2 + 2·1 respectively, clearly distinct. Carrying out the multiplications and additions numerically, these give the values 13 and 12 respectively. These sums are an additional discriminator that can be used to help count the distinct meeting structures. On this basis, we can distinguish 91 meeting structures for $n = 6$. However, there are actually M(6) = 92 distinct meeting structures. The two that are indistinguishable based on the criteria we've described so far are ab + bc + de + df + ef and ab + bc + cd + de + ef. These have the same set of valences, and the same pairings of valences, but they are nevertheless distinct structures, as shown by the fact that the first can be reduced to ab + bc + def, whereas the second is irreducible.

We can define another discriminator by replacing each letter not with the valence of that letter, but with the sum of the valences of all the letters with which it is paired. For example, in the first structure e is paired with d and with f, and these two letters appear twice each, for a total of 4, so we replace the letter e with 4. Doing the same for all the other letters in these structures, and carrying out the multiplications and additions numerically, we get

$$2.2 + 2.2 + 4.4 + 4.4 + 4.4 = 56$$
$$2.3 + 3.4 + 4.4 + 4.3 + 3.2 = 52$$

So this criterion enables us to distinguish between these two structures. On the same basis we can determine that M(7) = 369, meaning that there are 369 distinct meeting structures for n = 7 segments. However, for n = 8 segments the criteria we have described so far are not sufficient to distinguish between all the different structures. We find that there are at least 1800 such structures by these criteria, but in fact there are 1807. It's interesting to try to identify the most efficient discriminators between these structures. Obviously if we are given two sets of pairing expressions we can simply try all possible permutations of the letters and terms, to see if they can be brought into agreement, but this in not feasible for large values of n.

It's worth noting that if $a_j a_k$ is a connected pair with $j < k$, then $a_j a_i$ is also a connected pair for all indices i between j and k. For example, if bf is a connected pair, then bc, bd, and be are also connected pairs. It is this requirement that reduces the number of possible structures from $2^{n(n-1)/2}$ down to n!.

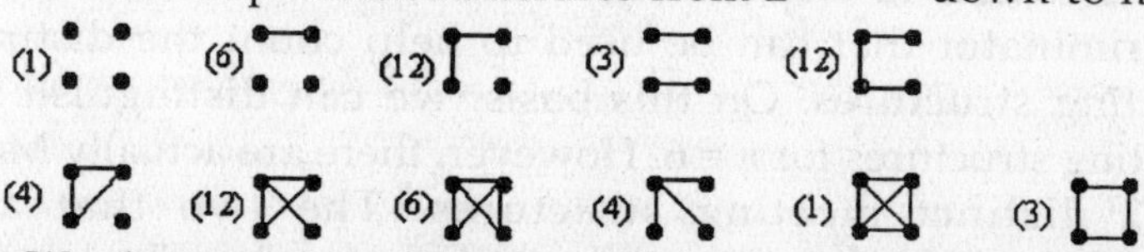

We then need to allow for equivalence under all possible permutations of the letters to reduce the number of distinct structures from n! down to M(n). These structures are often called "interval graphs", because they can be represented pictorially as a set of connections between the n nodes of a plane graph. Consider, for example, all $2^{4(4-1)/2}$ = 64 plane graphs involving four nodes, shown below.

The numbers in parentheses indicate how many of each of these structures there are if we regard the nodes as having distinct identities. (A convenient check is given by confirming that the total numbers of graphs with k edges is C(6, k).) If, on the other hand, the nodes are regarded as indistinguishable (as is the convention for graphs), then the eleven arrangements illustrated above represent all the distinct graphs on four nodes.

Ten of these graphs are interval graphs, corresponding to the ten "meeting structures" described previously. However, the "square cycle" graph shown on the right (in red) is *not* an interval graph, i.e., it does not correspond to any meeting structure of overlapping continuous intervals on a line. This stands to reason, because an interval graph cannot contain any minimal closed cycle of length greater than three.

One way of characterizing the unique meeting structures for n segments is in terms of the determinant of a certain n′n matrix. Let a_k with $k = 1, 2,.., n$ designate the n segments, and define an n′n matrix A with the non-zero components $A_{j,k} = 1$ for each pair of indices j, k such that $a_j a_k$ (or $a_k a_j$) is a connected pair. In addition, we set the diagonal values of A to the parameter x, so we have $A_{j,j} = x$ for $j = 1$ to n. (Note that a meeting structure is specified as a union of m pair-wise connections. If no two segments meet, then $m = 0$. If there is a connection between every pair of segments, then $m = n(n-1)/2$.)

Given the matrix A for a particular structure, the determinant of A is a polynomial in the parameter x. Since the determinant of a matrix is invariant under permutations of the row/columns, it follows that this polynomial characterizes a full equivalence class of meeting structures. Thus we can use these polynomials to discriminate between distinct structures. To illustrate, consider the following two meeting structures for $n = 7$ segments:

$$a_2a_3 + a_3a_4 + a_3a_5 + a_4a_5 + a_4a_6 + a_4a_7 + a_5a_6 + a_5a_7$$
$$a_2a_3 + a_2a_4 + a_2a_5 + a_2a_6 + a_3a_4 + a_3a_5 + a_5a_6 + a_5a_7$$

These structures have the same valence sets (i.e., they each have two characters that appear in four pairs, one that appears in three pairs, two that appear in two pairs, and one that appears in one pair), so in order to distinguish between them using the previous method we need to go to higher-order valences. However, these structures correspond to the matrices

$$\begin{bmatrix} x & 0 & 0 & 0 & 0 & 0 & 0 \\ 0 & x & 1 & 0 & 0 & 0 & 0 \\ 0 & 1 & x & 1 & 1 & 0 & 0 \\ 0 & 0 & 1 & x & 1 & 1 & 1 \\ 0 & 0 & 1 & 1 & x & 1 & 1 \\ 0 & 0 & 0 & 1 & 1 & x & 0 \\ 0 & 0 & 0 & 1 & 1 & 0 & x \end{bmatrix} \qquad \begin{bmatrix} x & 0 & 0 & 0 & 0 & 0 & 0 \\ 0 & x & 1 & 1 & 1 & 1 & 0 \\ 0 & 1 & x & 1 & 1 & 0 & 0 \\ 0 & 1 & 1 & x & 0 & 0 & 0 \\ 0 & 1 & 1 & 0 & x & 1 & 1 \\ 0 & 1 & 0 & 0 & 1 & x & 0 \\ 0 & 0 & 0 & 0 & 1 & 0 & x \end{bmatrix}$$

and the determinants of these matrices are

$$x^2(x-1)(x^4+x^3-7x^2-x+4)$$
$$x(x^6-8x^4+6x^3+7x^2-4x-1).$$

This shows how we can immediately distinguish between these two structures. (For automated numerical purposed, we can simply evaluate these polynomials for a few arbitrary values of x to discriminate between them.)

The matrices and characteristic polynomials for the four distinct meeting structures for $n = 3$ segments are shown below.

$$\det\begin{bmatrix} x & 0 & 0 \\ 0 & x & 0 \\ 0 & 0 & x \end{bmatrix} = x^3 \qquad \det\begin{bmatrix} x & 1 & 0 \\ 1 & x & 0 \\ 0 & 0 & x \end{bmatrix} = x(x^2-1)$$

$$\det\begin{bmatrix} x & 1 & 0 \\ 1 & x & 1 \\ 0 & 1 & x \end{bmatrix} = x(x^2-2) \qquad \det\begin{bmatrix} x & 1 & 1 \\ 1 & x & 1 \\ 1 & 1 & x \end{bmatrix} = x(x-1)^2(x+2)$$

Likewise we can determine the characteristic polynomials for each of the 10 interval graphs on $n = 4$ nodes, shown below.

∅	x^4
ab	$x^2(x-1)(x+1)$
ab+bc	$x(x-1)^2$
ab+cd	$(x-1)^2(x+1)^2$
abc	$x(x+2)(x-1)^2$
ab+bc+cd	$(x^2+x-1)(x^2-x-1)$

abc+cd	$(x-1)(x^3+x^2-3x-1)$
abc+cd	$x(x-1)(x^2+x-4)$
abc+abd	$x^2(x^2-3)$
ab+ac+ad	$(x+3)(x-1)^3$

As noted previously, the only other plane graph on four nodes is the four-loop

ab + bc + cd + da $x^2(x-2)(x+2)$

MEETING PROBABILITIES

Given n independent random variables, each evenly distributed over the interval 0 to 1, the probability that all n are within q of each other (for any $q < 1$) is

$$p_n = nq^{n-1} - (n-1)q^n$$

This can be derived in several different ways. For example, we can divide the unit interval into k equal segments, and note that the probability of n randomly selected points all falling within j consecutive segments corresponds approximately to the probability that all n points fall within $q = (j/k)$ of each other.

This correspondence becomes exact in the limit as k goes to infinity (holding q constant).

Equation (1) can also be derived from a geometrical point of view. Given a unit "cube" in n dimensions, equation (1) represents the fraction of the cube's content ("volume") consisting of points with orthogonal coordinates $[x_1, x_2, \ldots, x_n]$ such that $|x_i - x_j| < q$ for all i, j.

One possible generalization of this is to allow different tolerances on the different events. For example, suppose each of n people are to arrive at a certain location at some randomly chosen time between 1:00 PM and 2:00 PM, and each person will wait a certain amount of time before leaving.

Say, for example, with $n = 2$ people, one can wait for w_1 and the other can wait for w_2 (both expressed as fractions of the total time interval).

What is the probability that they will meet? Geometrically it's easy to see that this is just given by the are of the shaded region in the unit square shown below:

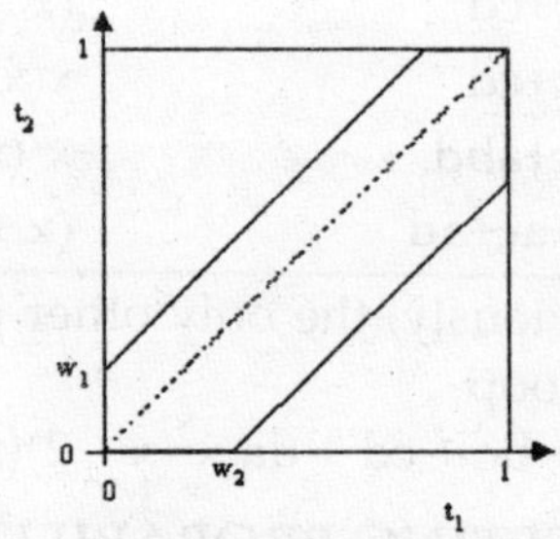

The shaded area equals to total square area minus the two excluded triangles, so we have

$$P_2 = 1 - \frac{(1-w_1)^2}{2} - \frac{(1-w_2)^2}{2} = w_1 + w_2 - \frac{1}{2}w_1^2 - \frac{1}{2}w_2^2$$

More generally, for n people with waiting times $w_1, w_2, .., w_n$, we just need to evaluate the integral of 1 over the following ranges of the arrival times $t_1, t_2, .., t_n$:

variable	lower limit	*upper* limit
t_1	0	1
t_2	max $(0, t_1 - w_2)$	max $(1, t_1 + w_1)$
t_3	max $(0, t_1 - w_3, t_2 - w_3)$	max $(1, t_1 + w_1 - t_2 + w_2)$
t_4	max $(0, t_1 - w_4, t_2 - w_3 - w_4)$	max $(1, t_1 + w_1, t_2 + w_2, t_3 + w_3)$
	etc	

Obviously this becomes extremely complicated as n increases. To illustrate, consider the case $n = 3$. We need to divide the problem into 12 separate double integrals over t_1 and t_2 in order to specify fixed integration limits for each one. Letting w_1 denote the smallest waiting time, and w_3 the largest, these 12 regions are illustrated below as the regions denoted by A through L.

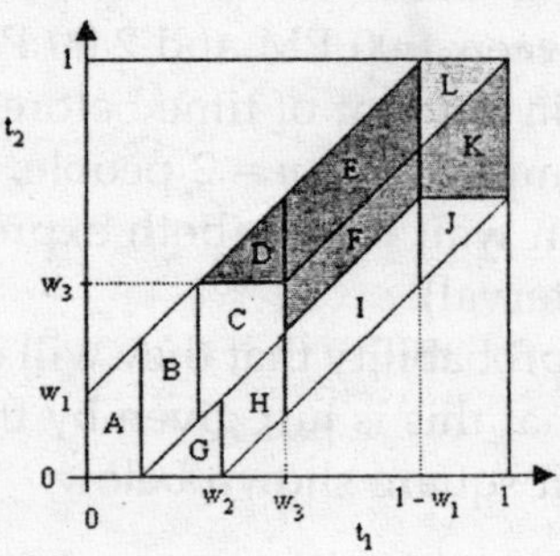

Within each of these regions we need to integrate the difference between the high and low integration limits for t_3. The integration limits and integrands are listed below

Region	t_1 Limits		t_2 Limits		Integrand
A	0	$w_2 - w_1$	0	$t_1 + w_1$	$t_1 + w_1$
B	$w_2 - w_1$	$w_3 - w_1$	$t_1 - w_2 + w_1$	$t_1 + w_1$	$t_1 + w_1$
C	$w_3 - w_1$	w_3	$t_1 - w_2 + w_1$	w_3	$t_1 + w_1$
D	$w_3 - w_1$	w_3	w_3	$t_1 + w_1$	$(t_1 + w_1) - (t_2 - w_3)$
E	w_3	$1 - w_1$	t_1	$t_1 + w_1$	$(t_1 + w_1) - (t_2 - w_3)$
F	w_3	$1 - w_1$	$t_1 - w_2 + w_1$	t_1	$(t_1 + w_1) - (t_1 - w_3)$
G	$w_2 - w_1$	w_2	0	$t_1 - w_2 + w_1$	$t_2 + w_2$
H	w_2	w_3	$t_1 - w_2$	$t_1 - w_2 + w_1$	$t_2 + w_2$
I	w_2	$1 - w_1$	$t_1 - w_2$	$t_1 - w_2 + w_1$	$(t_2 + w_2) - (t_1 + w_3)$
J	$1 - w_1$	1	$t_1 - w_2$	$t_1 - w_2$	$(t_2 + w_2) - (t_1 + w_3)$
K	$1 - w_1$	1	$1 - w_2$	t_1	$1 - (t_1 + w_3)$
L	$1 - w_1$	1	t_1	1	$1 - (t_1 + w_3)$

Evaluating all twelve of these integrals and simplifying gives the result

$$P_3 = w_1 w_2 + w_1 w_3 + w_2 w_3 - \frac{1}{2}\left(w_1 {w_2}^2 + w_1 {w_3}^2 + w_2 {w_3}^2\right)$$

$$-\frac{1}{3}{w_1}^2 - \frac{1}{6}{w_2}^3$$

For example, with 3 people who can wait $w_1 = 1/4$, $w_2 = 1/3$, and $w_3 = 1/2$ (15 minutes, 20 minutes, and 30 minutes, respectively) the probability of all three of them being at the location simultaneously during the hour is exactly 1435/5184, which is about 0.27681.

Naturally both the formulas for $n = 2$ and $n = 3$ reduce to equation when all the waiting times are equal. It's also worth noting that the formula for $n = 3$ is not symmetrical in the three waiting times, because the boundaries of the 3-dimensional solid representing the meeting region in phase space depend on a set of min and max selections, so the result depends on the ordering of the parameters, with w_1 the smallest and w_3 the longest waiting time. As n increases the direct integration method for arbitrary and unequal waiting times rapidly becomes much more complicated, because the waiting times must be truncated whenever someone arrives with less than his waiting time remaining before the end of the interval, and a similar truncation occurs at the beginning of the hour. Geometrically, we're dealing with the volume of an extremely complex n-dimensional asymmetrical polyhedron whose volume can't generally be given by a simple formula, except in the highly symmetrical case when all the d values are equal, or when n is very small (like the cases of $n = 2$ and $n = 3$ given above).

To derive a more efficient method of determining the probabilities for larger values of n, we may begin by considering a similar questions whose answers can immediately be given by fairly simple formulas with arbitrary n. Suppose we eliminate the boundary truncation in the original problem by wrapping the 1-hour interval around in a circle of unit circumference, and stipulating that the jth person is present for a continuous span of exactly w_j on that interval, and the location of each w_j is uniformly and independently distributed around the circle. Now we can ask for the probability that n spans of length $w_1, w_2, \ldots, w_n$ will have some common overlap. Let's also assume that the d values are all fairly small relative to the entire interval, so that $w_1 + w_2 + \ldots + w_n$ is less than 1. (This allows us to avoid complications due to wrap-around.)

Consider first the discrete case, i.e., suppose the circumference of the circle is divided into N equal increments, and the n randomly placed spans have discrete lengths of $m_1, m_2, \ldots, m_n$ increments respectively. The total number of possible arrangements of the spans is N^n, so we need to determine how many of those contain at least one increment of overlap between all the spans.

Obviously each configuration with overlap can be translated into N different positions, so we really just need to find the number of distinct intrinsic overlap configurations of these n spans.

If we fix one particular increment on the circle and require that all the spans contain that increment, then the number of arrangements that satisfy this requirement is obviously the product $m_1 m_2 ... m_n$. However, this counts the intrinsic arrangements with two increments of overlap twice, because of translation. Likewise it counts the arrangements with three increments of overlap three times, and so on. Thus the product of lengths represents an overestimation of the number of intrinsic arrangements with overlap.

But suppose we fix *two* particular consecutive increments on the circle and require that all the spans contain both of them. The number of arrangements that satisfy this requirement is obviously the product $(m_1-1)(m_2-1)...(m_n-1)$. This counts the arrangements with two increments of overlap only once, and it counts the arrangements with three increments of overlap twice, and so on. Therefore, if we subtract this product from the previous product, the result counts the number of arrangements with any number of increments of overlap exactly once. Hence the number of such arrangements is

$$m_1 m_2 ... m_n - (m_1 - 1)(m_2 - 1)....(m_n - 1)$$

It follows that the probability of overlap for these n spans placed randomly around the unit circle is

$$P_n = \frac{N\left[m_1 m_2 ... m_n - (m_1 - 1)(m_2 - 1)(m_n - 1)\right]}{N^n}$$

For example, with $n = 3$ the result is

$$P_3 = \frac{1 - m_1 - m_2 - m_3 + m_1 m_2 + m_1 m_3 + m_2 m_3}{N^2}$$

Obviously if we hold the ratios $w_1 = m_1/\text{N}$, $w_2 = m_2/N$, and $w_3 = m_3/\text{N}$ constant and increase N, all the terms in the numerator of degree less than two drop out, and we're left with the result for the continuous case

$$P_3 = w_1 w_2 + w_1 w_3 + w_2 w_3 = w_1 w_2 w_3 \left(\frac{1}{w_1} + \frac{1}{w_2} + \frac{1}{w_3}\right)$$

In general for n spans of lengths $w_1, w_2, .., w_n$ randomly distributed uniformly around a circle of unit circumference, with the condition that the sum $w_1 + w_2 + ... + w_n$ is less than 1, the probability of some common overlap among all n of the spans is

$$P_n = w_1 w_2 \ldots w_n \left(\frac{1}{w_1} + \frac{1}{w_2} + \ldots + \frac{1}{w_n} \right)$$

Now let's return to our original problem, to find the probability of meeting between n people with distinct waiting times in a fixed interval. Again we can start with the discrete case, and then go over to the continuous case at the end. We divide the unit interval into N segments, and let the integer m_j for $j = 1,2,..,n$ denote the waiting times (corresponding to Nw_j), arranged so that $1 \le m_1 \le m_2 \le \ldots \le m_n \le N$. We apply the same reasoning as in the case of the circle, except that now each fixed segment of overlap must be treated separately, because the segments near the starting time are truncated by the boundary. For example, the first waiting interval m_1 has only one position that includes the first segment, and it has only two positions that overlap with the second segment, and so on.

Only when we reach segments more than m_j from the boundary are there m_j intersecting positions. In general, the number of positions of the kth waiting interval that intersect with the jth segment is $\min(j, m_k)$. Hence, instead of having N times the product of the m_k values, we must evaluate the sum over j from 1 to N of the products of the $\min(j, m_k)$ values. Of course, as in the case of the circle, this represents an over-estimate, because if counts each configuration multiple times.As before, this is corrected by subtracting the products of the quantities $[\min(j, m_k)-1]$.

Consequently, the probability of meeting for n people with waiting times $m_1, m_2, \ldots, m_n$ is given by

$$P_n = \frac{1}{N^n} \sum_{j=1}^{N} \left\{ \left(\prod_{i=1}^{n} \min(j, m_i) \right) - \left(\prod_{i=1}^{n} \min(j, m_i) - 1 \right) \right\}$$

After expanding the products, the terms of degree n cancel, leaving only terms of degree $n-1$ and lower. Bringing $1/N^{n-1}$ in from the leading factor, and letting w_j and t denote the ratios m_j/N and j/N respectively, the resulting expression is purely a function of the w_j and the variable t, plus terms divided by some power of N. The latter vanish in the limit as

N goes to infinity, so only the terms of degree n–1 remain. (These are comprised of the n products of n–1 elements of the set min(j, m_j), j = 1,2,..,n.) In the continuous limit, the summation over the index j can be evaluated as an integration over the variable t. Noting that dj = Ndt, the remaining power of 1/N in the leading factor is cancelled, and we have in the continuous limit

$$P_n = \int_{t=0}^{1} s_{n-1}(\min\ (t, w_1), \min\ (t, w_1), ..., \min(t, w_n))\, dt$$

where $s_{n-1}(x_1, x_2,..,x_n)$ denotes the sum of all products of n–1 arguments. Splitting up this integral into the ranges from 0 to w_1, from w_1 to w_2, and so on, we can eliminate the min functions and write this as a sum of elementary integrals. To simplify the expressions, it's convenient to define the following symbols for the kth and (k–1)th symmetric functions of the k smallest waiting times:

$$\pi_k = w_1 w_2 ... w_1 \quad \sigma_k = \left(\frac{1}{w_1} + \frac{1}{w_2} + ... \frac{1}{w_k}\right)\pi_k$$

with p_0 = 1 and s_0 = 0. Also, let w_0 = 0 and w_{n+1} = 1. In terms of these parameters we can write

$$P_n = \sum_{j=1}^{n} \left[\int_{w_j}^{w_{j+1}} \left\{ (n-j) p_j t^{n-1-j} + \sigma_j t^{n-j} \right\} dt \right]$$

Evaluating the integrals, this becomes

$$P_n = \sum_{j=1}^{n-1} \left[\left(w_{j+1}{}^{n-j} - w_j{}^{n-j} \right) \pi_j \right]$$

$$+ \sum_{j=1}^{n} \left[\left(w_{j+1}{}^{n-j+1} - w_j{}^{n-j+1} \right) \frac{\sigma_j}{n-j+1} \right]$$

In view of the identity $p_{j+1} = p_j\, w_{j+1}$, the argument of the first summation can be written as

$$\left(w_{j+1}{}^{n-j} - w_j{}^{n-j} \right) \pi_j = \pi_{j+1} w_{j+1}{}^{n-[j+1]} - \pi_j w_j{}^{n-j}$$

Hence these terms constitute a telescoping sequence, with the net sum π_n. Therefore, we have the result

$$P_n = \pi_n + \sum_{j=1}^{n} \left[\left(w_{j+1}^{\ n-j+1} - w_j^{\ n-j+1} \right) \frac{\sigma_j}{n-j+1} \right]$$

Notice that this formula reduces to equation (1) if all the waiting times are equal. Also, it's easily verified that this formula gives the results previously derived for $n = 2$ and $n = 3$. For another illustration, consider the case $n = 4$. Using this expression, the probability of all parties meeting is

$$P_4 = \pi_4 + \frac{w_2^{\ 4} - w_1^{\ 4}}{4}\sigma_1 + \frac{w_3^{\ 3} - w_2^{\ 3}}{3}\sigma_2 + \frac{w_4^{\ 4} - w_3^{\ 4}}{2}\sigma_3 + \frac{1 - w_4}{4}\sigma_4$$

Inserting the explicit expressions for the symmetric functions, and simplifying, this gives the formula

$$P_4 = w_1 w_2 w_3 + w_1 w_2 w_4 + w_1 w_3 w_4 + w_2 w_3 w_4$$
$$- \frac{w_4^{\ 2}}{2}(w_1 w_2 + w_1 w_3 + w_2 w_3) - \frac{w_3^{\ 3}}{6}(w_1 + w_2)$$
$$- \frac{w_1 w_2 w_3^{\ 2}}{2} - \frac{w_1 w_2^{\ 2}}{3} - \frac{w_1^{\ 4}}{4} - \frac{w_2^{\ 4}}{12}$$

To give the explicit formula more directly, the summation in equation (2) can be made more explicit if we collect terms by w_j. This gives

$$P_n = \pi_n - \frac{w_1^{\ n}}{n}\sigma_1 + \sum_{j=1}^{n-1} \left[\frac{w_{j+1}^{\ n-j+1}}{n-j+1}\sigma_j - \frac{w_{j+1}^{\ n-j}}{n-j}\sigma_{j+1} \right] + \sigma_n$$

Making use of the identity $s_{j+1} = w_{j+1}\, s_j + p_j$, the summation splits into a sum over the p_j and a sum over the s_j as follows

$$P_n = \pi_n - \frac{w_1^{\ n}}{n}\sigma_1 + \sum_{j=1}^{n-1} \left[\frac{w_{j+1}^{\ n-j}}{n-j}\pi_j \right] - \sum_{j=1}^{n-1} \left[\frac{w_{j+1}^{\ n-j+1}}{(n-j)(n-j+1)}\sigma_j \right] + \sigma_n$$

The argument of the first summation when $j = n-1$ is simply $w_n \pi_{n-1} = \pi_n$, so this cancels with the leading p_n term.

Also, since $\sigma_1 = \pi_0 = 1$, we can bring the term involving w_1 into the first summation, shift the indices, and combine one power of w_j with π_{j-1}, to give the result

$$P_n = \sigma_n - \sum_{j=1}^{n-1}\left[\frac{w_j^{\,n-j}}{n-j+1}\pi_j\right] - \sum_{j=1}^{n-1}\left[\frac{w_{j+1}^{\,n-j+1}}{(n-j)(n-j+1)}\sigma_j\right]$$

(For comparison, recall that the circular case gave simply $P_n = \sigma_n$.) Each s_j represents j terms, so the total number of terms (i.e., individual products) in P_n is

$$n + (n+1) + \frac{(n-1)n}{2} = \frac{n^2 + 3n - 2}{2}$$

Chapter 6

Graphs

DEFINITION

Graph theory is a branch of mathematics concerned about how networks can be encoded and their properties measured. In transport geography most networks have an obvious spatial foundation, namely road, transit and rail networks, which tend to be defined more by their links than by their nodes. This it is not necessarily the case for all transportation networks. For instance, maritime and air networks tend to be more defined more by their nodes than by their links since links are often not clearly defined.

A telecommunication system can also be represented as a network, while its spatial expression can have limited importance and would actually be difficult to represent. Mobile telephone networks or the Internet, possibly to most complex graphs to be considered, are relevant cases of networks having a structure that can be difficult to symbolize. However, cellular phones and antennas can be represented as nodes while the links could be individual phone calls. Servers, the core of the Internet, can also be represented as nodes within a graph while the physical infrastructure between them, namely fibre optic cables, can act as links. Consequently, all transport networks can be represented by graph theory in one way or the other.

FUNDAMENTAL OF GRAPH THEORY

Graph. A graph G is a set of vertex (nodes) v connected by edges (links) e. Thus $G=(v, e)$.

Vertex (Node): A node v is a terminal point or an intersection point of a graph. It is the abstraction of a location such as a city, an administrative division, a road intersection or a transport terminal (stations, terminuses, harbors and airports).

Edge (Link): An edge e is a link between two nodes. The link (i, j) is of initial extremity i and of terminal extremity j. A link is the abstraction of a transport infrastructure supporting movements between nodes. It has a direction that is commonly represented as an arrow. When an arrow is not used, it is assumed the link is bi-directional.

Sub-Graph: A sub-graph is a subset of a graph G where p is the number of sub-graphs. For instance $G' = (v', e')$ can be a distinct sub-graph of G. Unless the global transport system is considered in its whole, every transport network is in theory a sub-graph of another. For instance, the road transportation network of a city is a sub-graph of a regional transportation network, which is itself a sub-graph of a national transportation network.

Buckle: A link that makes a node correspond to itself is a buckle.

Planar Graph: A graph where all the intersections of two edges are a vertex. Since this graph is located within a plane, its topology is two-dimensional.

Non-planar Graph: A graph where there are no vertex at the intersection of at least two edges. This implies a third dimension in the topology of the graph since there is the possibility of having a movement "passing over" another movement such as for air transport. A non-planar graph has potentially much more links than a planar graph.

LINKS AND THEIR STRUCTURES

A transportation network enables flows of people, freight or information, which are occurring along its links. Graph theory must thus offer the possibility of representing movements as linkages, which can be considered over several aspects:

Connection: A set of two nodes as every node is linked to

the other. Considers if a movement between two nodes is possible, whatever its direction. Knowing connections makes it possible to find if it is possible to reach a node from another node within a graph.

Path: A sequence of links that are traveled in the same direction. For a path to exist between two nodes, it must be possible to travel an uninterrupted sequence of links. Finding all the possible paths in a graph is a fundamental attribute in measuring accessibility and traffic flows.

Chain: A sequence of links having a connection in common with the other. Direction does not matter.

Length of a Link, Connection or Path. Refers to the label associated with a link, a connection or a path. This label can be distance, the amount of traffic, the capacity or any attribute of that link. The length of a path is the number of links (or connections) in this path.

Cycle: Refers to a chain where the initial and terminal node is the same and that does not use the same link more than once is a cycle.

Circuit: A path where the initial and terminal node corresponds. It is a cycle where all the links are traveled in the same direction. Circuits are very important in transportation because several distribution systems are using circuits to cover as much territory as possible in one direction (delivery route).

PROPERTIES

The organization of nodes and links in a graph conveys a structure that can be described and labeled. The basic structural properties of a graph are:

Symmetry and Asymmetry. A graph is symmetrical if each pair of nodes linked in one direction is also linked in the other. By convention, a line without an arrow represents a link where it is possible to move in both directions.

However, both directions have to be defined in the graph. Most transport systems are symmetrical but asymmetry can often occur as it is the case for maritime (pendulum) and air services.

Asymmetry is rare on road transportation networks, unless one-way streets are considered.

Completeness: A graph is complete if two nodes are linked in at least one direction. A complete graph has no sub-graph.

Connectivity: A complete graph is described as connected if for all its distinct pairs of nodes there is a linking chain. Direction does not have importance for a graph to be connected, but may be a factor for the *level* of connectivity. If p>1 the graph is not connected because it has more than one sub-graph. There are various levels of connectivity, depending on the degree at which each pair of nodes is connected.

Complementary: Two sub graphs are complementary if their union results in a complete graph. Multimodal transportation networks are complementary as each sub-graph benefits from the connectivity of other sub-graphs.

Root: A node r where every other node is the extremity of a path coming from r is a root. Direction has an importance. A root is generally the starting point of a distribution system, such as a factory or a warehouse.

Trees: A connected graph without a cycle is a tree. A tree has the same number of links than nodes plus one. ($e = v\text{-}1$). If a link is removed, the graph ceases to be connected. If a new link between two nodes is provided, a cycle is created. A branch of root r is a tree where no links are connecting any node more than once.

Articulation Node: In a connected graph, a node is an articulation node if the sub-graph obtained by removing this node is no longer connected. It therefore contains more than one sub-graph ($p > 1$). An articulation node is generally a port or an airport, or an important hub of a transportation network, which serves as a bottleneck.

Isthmus: In a connected graph, an isthmus is a link that is creating, when removed, two sub-graphs having at least one connection.

HISTORY

The origin of graph theory can be traced back to Euler's work on the Konigsberg bridges problem, which subsequently led to the concept of an Eulerian graph. The study of cycles on polyhedra by the Thomas P. Kirkman and William R. Hamilton led to the concept of a Hamiltonian graph.

The concept of a tree, a connected graph without cycles, appeared implicitly in the work of Gustav Kirchhoff, who employed graph-theoretical ideas in the calculation of currents in electrical networks or circuits. Later, Arthur Cayley, James J. Sylvester George Polya, and others use 'tree' to enumerate chemical molecules. The study of planar graphs originated in two recreational problems involving the complete graph K_5 and the complete bipartite graph $K_{3,3}$. These graphs proved to be planarity, as was subsequently demonstrated by Kuratowski. First problem was presented by A. F. Mobius around the year 1840 as follows:

> Once upon a time, there was a king with five sons.
>
> In his will he stated that after his death the sons should divide the kingdom into five provinces so that the boundary of each province should have a frontiers line in common with each of the other four provinces.

Here the problem is whether one can draw five mutually neighboring regions in the plane.

> The king further stated that all five brothers should join the provincial capital by roads so that no two roads intersect.

Here the problem is that deciding whether the graph K_5 is planar.

The origin of second problem is unknown but it is first mentioned by H. Dudeney in 1913 in its present form.

> The puzzle is to lay a water, gas, and electricity to each of the three houses without any pipe crossing another.

This problem is that of deciding whether the graph $K_{3,3}$ is planar.

EXAMPLES

Informally, a graph is a diagram consisting of points, called vertices, joined together by lines, called edges; each edge joins exactly two vertices. A graph G is a triple consisting of a vertex set of V(*G*), an edge set E(G), and a relation that associates with each edge two vertices (not necessarily distinct) called its endpoints.

TYPES OF GRAPHS

GRAPH

A graph *G* = (*V*, *E*) consists of a (finite) set denoted by *V*, or by *V*(*G*) if one wishes to make clear which graph is under consideration, and a collection *E*, or *E*(*G*), of unordered pairs $\{u, v\}$ of distinct elements from V. Each element of V is called a vertex or a point or a node, and each element of E is called an edge or a line or a link.

Formally, a graph G is an ordered pair of dsjoint sets (V, E), where $E \subseteq V \times V$. Set V is called the vertex or node set, while set E is the edge set of graph G. Typically, it is assumed that self-loops (i.e. edges of the form (u, u), for some $u \in V$) are not contained in a graph.

DIRECTED AND UNDIRECTED GRAPH

A graph G = (V, E) is directed if the edge set is composed of ordered vertex (node) pairs. A graph is undirected if the edge set is composed of unordered vertex pair.

Vertex Cardinality

The number of vertices, the cardinality of V, is called the order of graph and devoted by $|V|$. We usually use *n* to denote the order of G. The number of edges, the cardinality of E, is called the size of graph and denoted by $|E|$. We usually use *m* to denote the size of G.

Neighbour Vertex and Neighborhood

We write $v_i v_j \in E(G)$ to mean $\{v_i, v_j\} \in E(G)$, and if $e = v_i v_j \in E(G)$, we say v_i and v_j are adjacent.

Formally, given a graph G = (V, E), two vertices $v_i, v_j \in V$ are said to be neighbors, or adjacent nodes, if $(v_i, v_j) \in E$. If G is directed, we distinguish between incoming neighbors of v_i

(those vertices $v_j \in V$ such that $(v_j, v_i) \in E$) and outgoing neighbors of v_i (those vertices $v_j \in V$ such that $(v_i, v_j) \in E$).

The open neighborhood N(v) of the vertex v consists of the set vertices adjacent to v, that is, $N(v) = \{w \in v: vw \in E\}$. The closed neighborhood of v is $N[v] = N(v) \cup \{v\}$. For a set S Í V, the open neighborhood N(S) is defined to be $U_{v \in S} N(v)$, and the closed neighborhood of S is $N[S] = N(S) \cup S$.

Vertex Degree: The degree deg(v) of vertex v is the number of edges incident on v or equivalently, deg(v) = |N(v)|. The degree sequence of graph is (deg(v_1), deg(v_2),..., deg(v_n)), typically written in nondecreasing or nonincreasing order. The minimum and maximum degree of vertices in V(G) are denoted by d(G) and ä(G), respectively. If d(G) = Ä(G) = r, then graph G is said to be regular of degree r, or simply r-regular.

Formally, given a graph G = (V, E), the degree of a vertex $v \in V$ is the number of its neighbors in the graph. That is,

$$\deg(v) = |\{u \in V: (v, w) \in E\}|.$$

If G is directed, we distinguish between in-degree (nimber of incoming neighbors) and out-degree (number of outgoing neighbors) of a vertex.

Loop and Multiple Edges: A loop is an edge whose endpoints are equal i.e., an edge joining a vertex to it self is called a loop. We say that the graph has multiple edges if in the graph two or more edges joining the same pair of vertices.

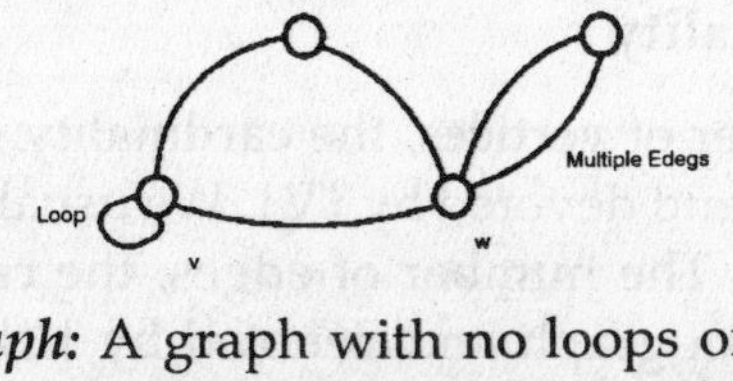

Simple Graph: A graph with no loops or multiple edges is called a simple graph. We specify a simple graph by its set of vertices and set of edges, treating the edge set as a set of unordered pairs of vertices and write $e = uv$ (or $e = vu$) for an edge e with endpoints u and v.

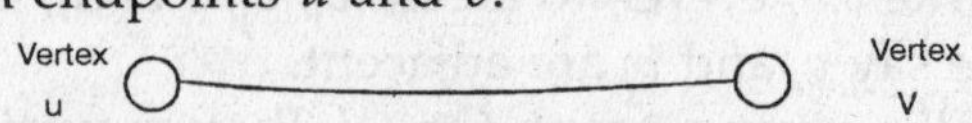

When u and v are endpoints of an edge, they are adjacent and are neighbors.

CONNECTED GRAPH

A graph that is in one piece is said to be connected, whereas one which splits into several pieces is disconnected.

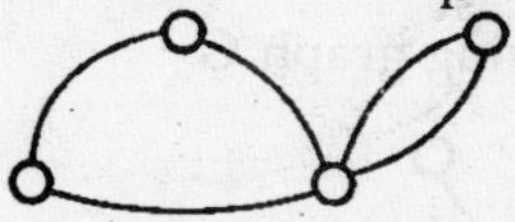

Connected Non-Simple Graph

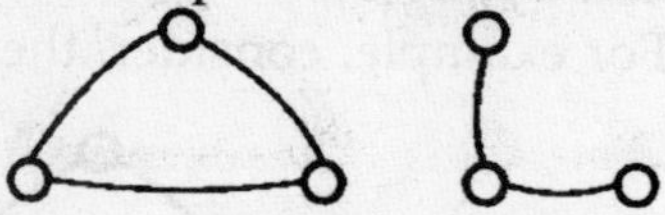

Disconnected Simple Graph

A graph G is connected if there is a path in G between any given pair of vertices, otherwise it is disconnected. Every disconnected graph can be split up into a number of connected subgraphs, called components.

SUBGRAPH

Let *G* be a graph with vertex set *V*(*G*) and edge-list E(G). A subgraph of *G* is a graph all of whose vertices belong to *V*(*G*) and all of whose edges belong to *E*(*G*). For example, if *G* is the connected graph below:

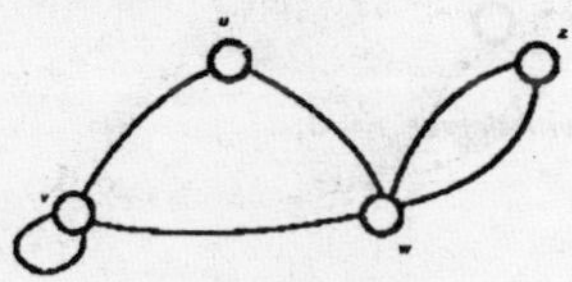

where $V(G) = \{u, v, w, z\}$ and $E(G) = (uv, uw, vv, vw, wz, wz\}$ then the following four graphs are subgraphs of G.

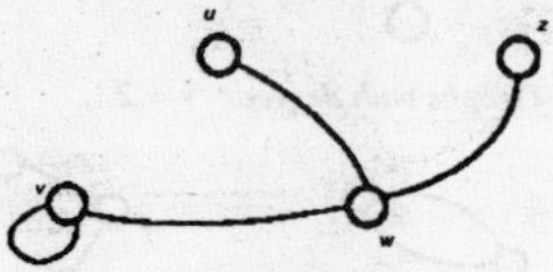

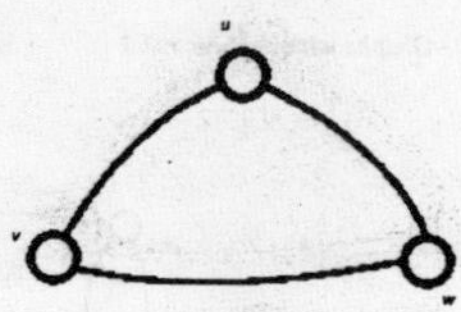

Degree (or Valency)

Let G be a graph with loops, and let v be a vertex of G. The degree of v is the number of edges meeting at v, and is denoted by $deg(v)$.

For example, consider, the following graph G

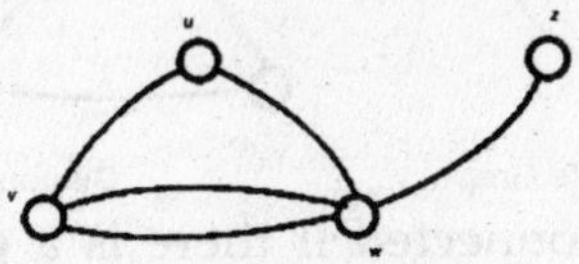

The graph G has $deg(u) = 2$, $deg(v) = 3$, $deg(\underline{w}) = 4$ and $deg(z) = 1$.

REGULAR GRAPH

A graph is regular if all the vertices of G have the same degree. In particular, if the degree of each vertex is r, the G is regular of degree r.

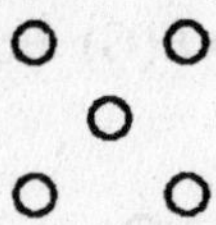

Graph with degree $r = 0$

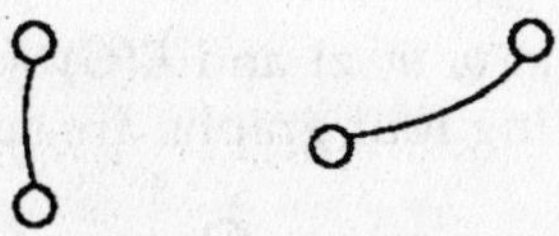

Graphs with degrees $r = 2$

Graphs with degree $r = 2$

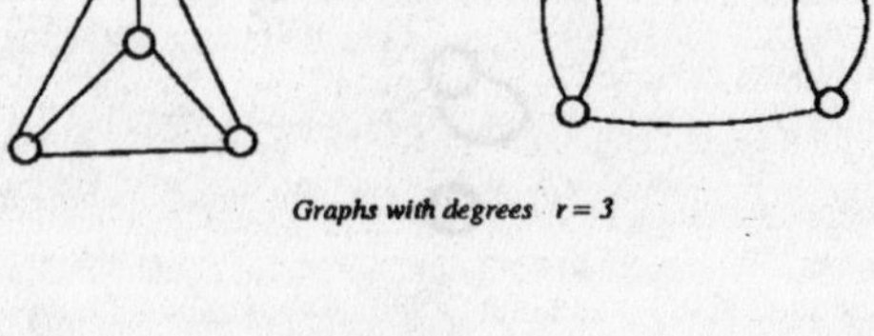

Graphs with degrees $r = 3$

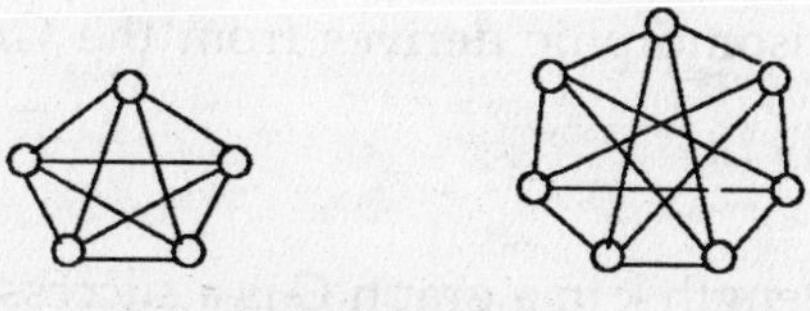

Graphs with degrees $r = 4$

The Handshaking Lemma: In any graph, the sum of all the vertex-degree is equal to twice the number of edges.

Proof: Since each edge has two ends, it must contribute exactly 2 to the sum of the degrees. The result follows immediately. The Following are the consequences of the Handshaking lemma.

1. In any graph, the sum of all the vertex-degree is an even number.
2. In any graph, the number of vertices of odd degree is even.
3. If G is a graph which has n vertices and is regular of degree r, then G has exactly 1/2 nr edges.

ISOMORPHIC GRAPHS

Two graph G and H are isomorphic if H can be obtained from G by relabeling the vertices – that is, if there is a one-to-one correspondence between the vertices of G and those of H, such that the number of edges joining any pair of vertices in G is equal to the number of edges joining the corresponding pair of vertices in H. For example, two unlabeled graphs, such as

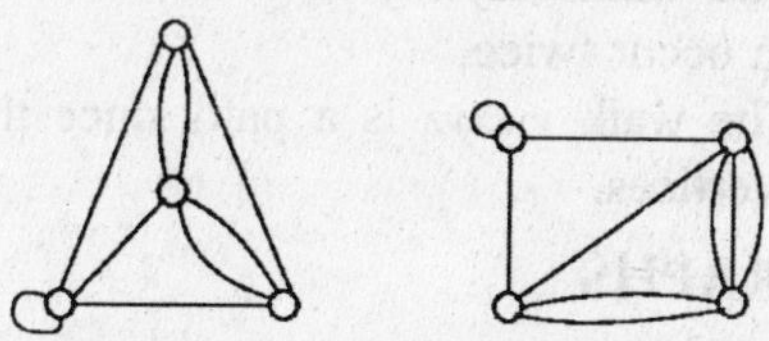

are isomorphic if labels can be attached to their vertices so that they become the same graph.

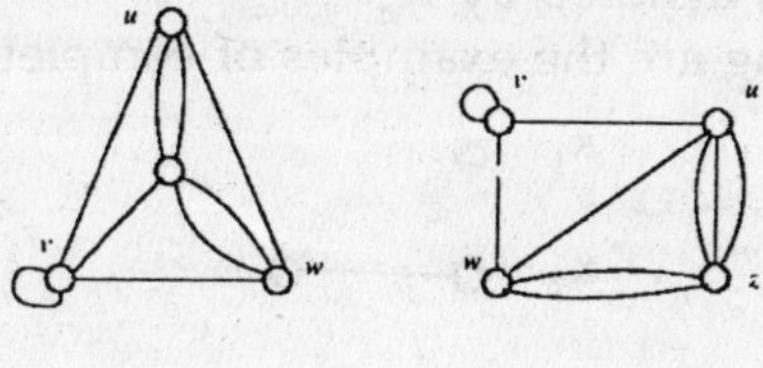

The word isomorphic derives from the Greek for same and form.

WALK

A walk of length k in a graph G is a succession of *k* edges of G of the form *uv, vw, wx,..., yz.*

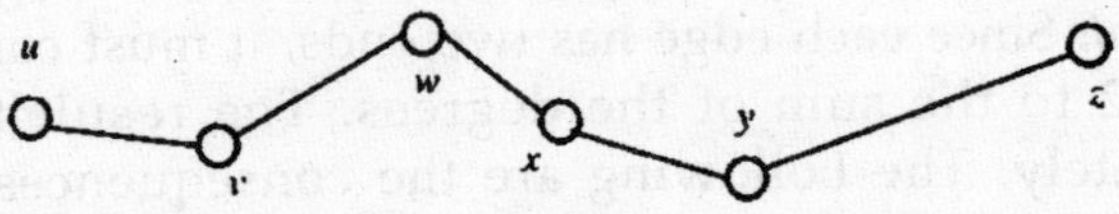

We denote this walk by *uvwx.. yz* and refer to it as a walk between *u* and *z.*

Trail and Path

If all the edges (but no necessarily all the vertices) of a walk are different, then the walk is called a trail. If, in addition, all the vertices are difficult, then the trail is called path.

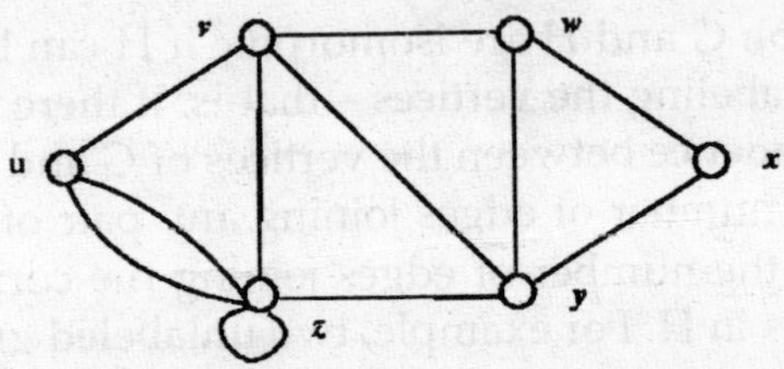

1. The walk *vzzywxy* is a trail since the vertices y and z both occur twice.
2. The walk *vwxyz* is a path since the walk has no repeated vertices.

COMPLETE GRAPHS

A computer graph is a graph in which every two distinct vertices are joined by exactly one edge. The complete graph with *n* vertices is denoted by K_n.

The following are the examples of complete graphs.

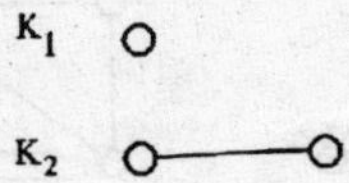

K_3
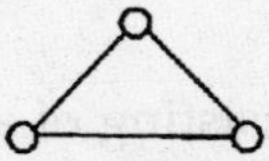

K_4
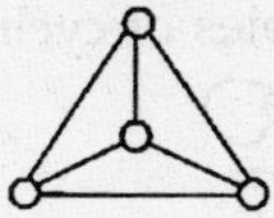
OR
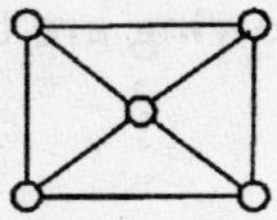

K_5
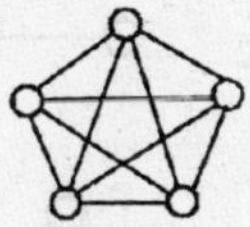

K_6
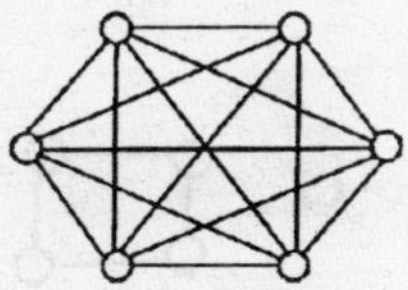

The graph K_n is regular of degree n-1, and therefore has 1/2n(n-1) edges, by consequence 3 of the handshaking lemma.

NULL GRAPHS

A null graphs is a graph containing no edges. The null graph with n vertices is denoted by N_n.

The following are the examples of null graphs.

Note that N_n is regular of degree 0.

N_1

N_2

N_3
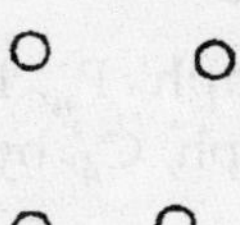

N_4

CYCLE GRAPHS

A cycle graph is a graph consisting of a single cycle. The cycle graph with n vertices is denoted by C_n.

The following are the examples of cyclic graphs.

C_1 

C_2 

C_3 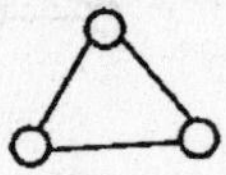

C_4 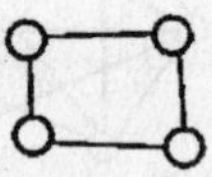

Note that C_n is regular of degree 2, and has n edges.

PATH GRAPHS

A path graph is a graph consisting of a single path. The path graph with n vertices is denoted by P_n.

The following are the examples of path graphs.

P_1 

P_2 

P_3 

P_4 

Note that path graph, P_n, has n-1 edges, and can be obtained from cycle graph, C_n, by removing any edge.

BIPARTITE GRAPHS

A bipartite graph is a graph whose vertex-set can be split

into two sets in such a way that each edge of the graph joins a vertex in first set to a vertex in second set. The examples of bipartite graphs are:

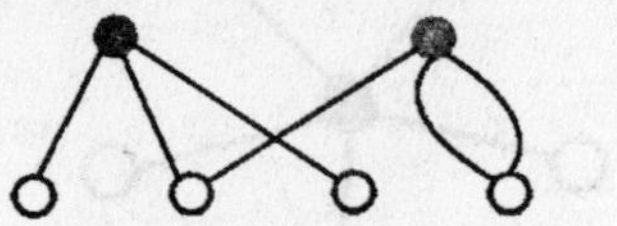

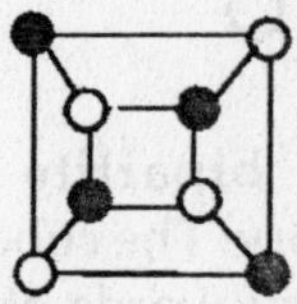

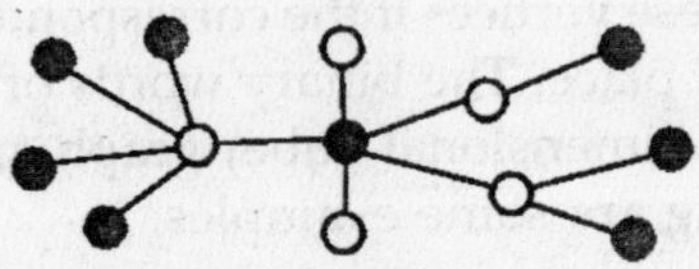

Complete Bipartite Graph

A complete bipartite graph is a bipartite graph in which each vertex in the first set is joined to each vertex in the second set by exactly one edge. The complete bipartite graph with *r* vertices and 3 vertices is denoted by $K_{r,s}$. The following are some examples.

$K_{2,2}$

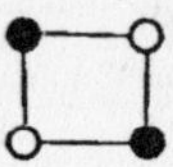

$K_{2,4}$

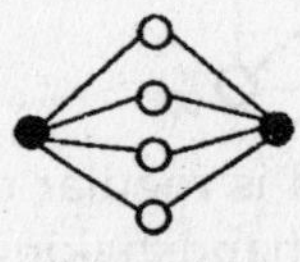

OR

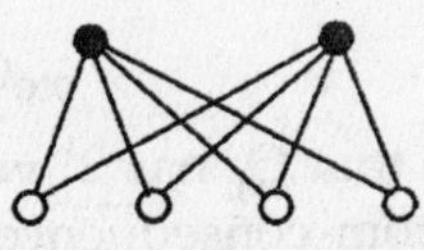

$K_{3,3}$

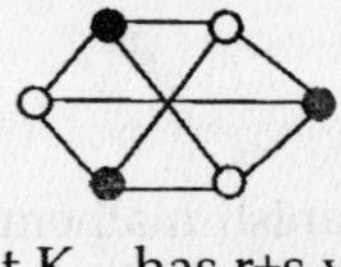

OR

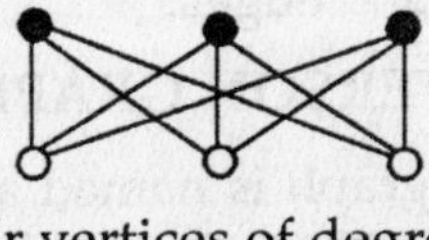

Note that $K_{r,s}$ has r+s vertices (r vertices of degrees, and s vertices of degree r), and rs edges. Note also that $K_{r,s} = K_{s,r}$.

An Important Note: A complete bipartite graph of the form $K_{r,s}$ is called a star graph.

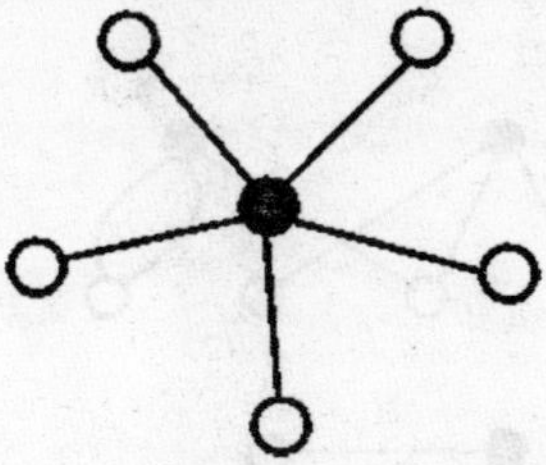

CUBE GRAPH

The cube graphs is a bipartite graphs and have appropriate in the coding theory. The cube graphs constructed by taking as vertices all binary words of a given length and joining two of these vertices if the corresponding binary words differ in just one place. The binary words of length *k* is called the *k*-cube (or *k*-dimensional cube) graph and is denoted by Q_k. The following are some examples.

Q_1

Q_2

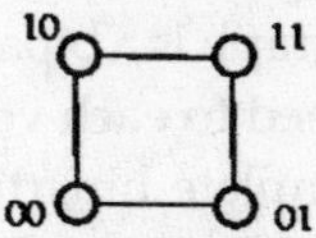

Q_3

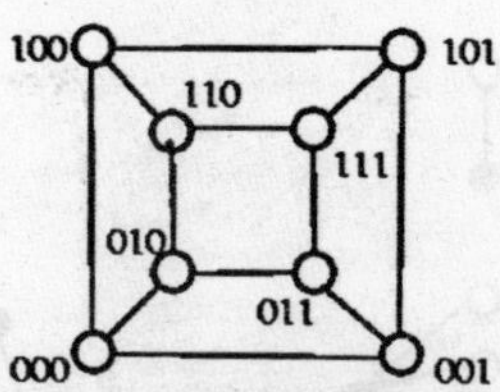

Note that Q_k has 2^k vertices and is regular of degree *k*. It follows from consequence 3 of the handshaking lemma that Q_k has $k^* 2^{k-1}$ edges.

THE PETERSON GRAPH

This graph is named after a Danish mathematician, Julius Peterson(1839-1910), who discovered the graph in a paper of 1898.

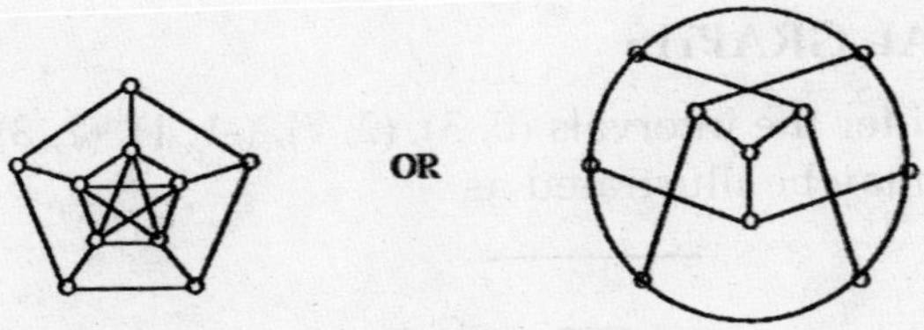

TREE GRAPH

A tree is a connected graph which has no cycles.

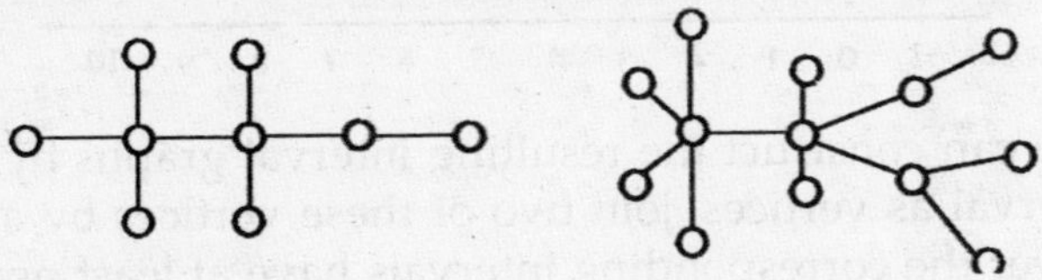

Spanning Tree

If *G* is a connected graph, the spanning tree in *G* is a subgraph of *G* which includes every vertex of *G* and is also a tree.

Consider the following graph

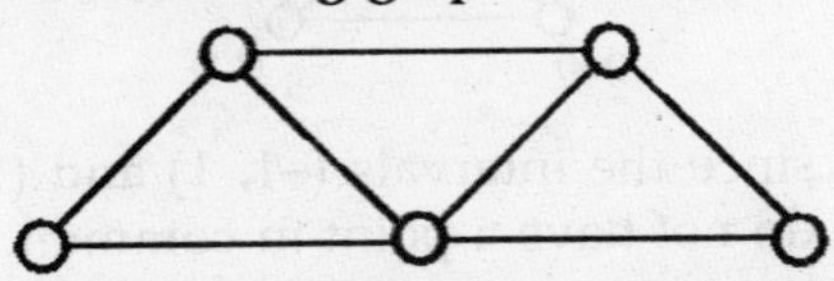

The following are the three of its spanning trees:

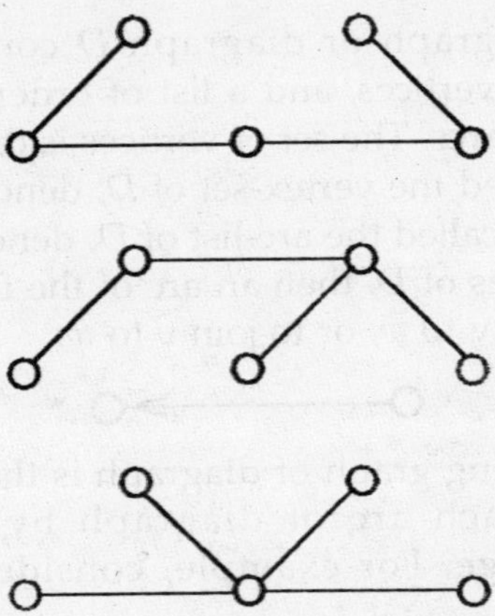

INTERVAL GRAPHS

Consider the intervals (0, 3), (2, 7), (-1, 1), (2, 3), (1, 4), (6, 8) which may be illustrated as

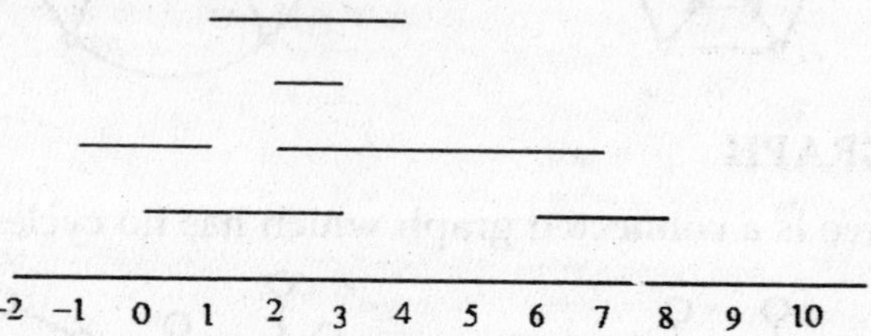

We can construct the resulting interval graphs by taking the interval as vertices, join two of these vertices by an edge whenever the corresponding intervals have at least one point in common.

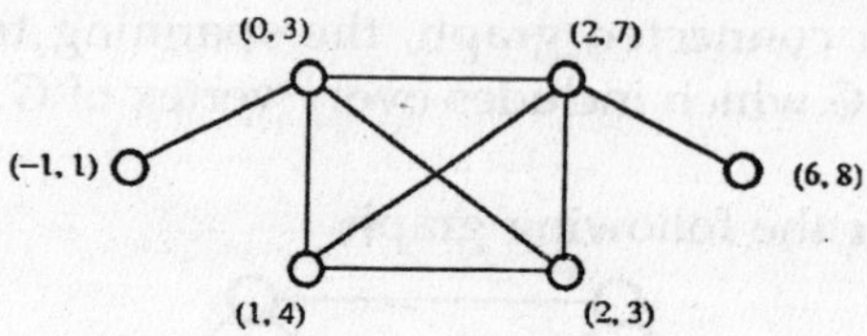

Note that since the intervals (–1, 1) and (1, 4) are open intervals, they do not have a point in common.

Diagraph

A directed graph or diagraph *D* consists of a set of elements, called vertices, and a list of ordered pairs of these elements, called arcs. The set of vertices is called arcs. The set of vertices is called the vertex-set of *D*, denoted by *V*(*D*), and the list of arcs is called the arc-list of *D*, denoted by *A*(*D*). If v and w are vertices of *D*, then an arc of the form *vw* is said to be directed from v to *w*, or to join *v* to *w*.

The underlying graph of diagraph is the graph obtained by replacing each arc of diagraph by corresponding (undirected) edge. For example, consider the following digraph

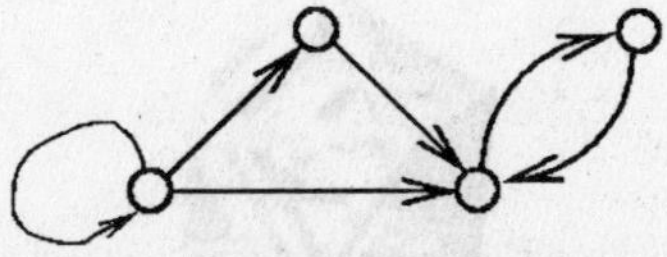

The underlying graph of the above digraph is

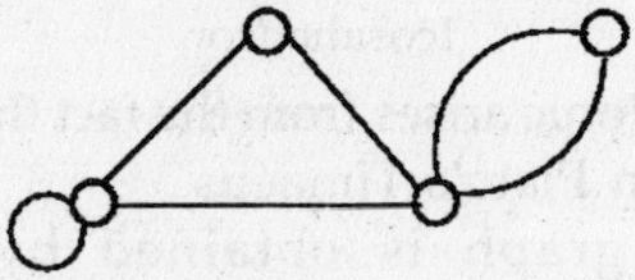

THE PLATONIC GRAPHS

The following regular solids are called the Platonic solids:

Tetrahedron

Hexahedron (cube)

Octahedron

Dodecahedron

Icosahedron

The name Platonic arises from the fact that these five solids were mentioned in Plato's Timaeus.

A Platonic graph is obtained by projecting the corresponding solid on to a plane.

Tetrahedron

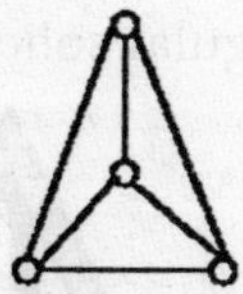

Cube

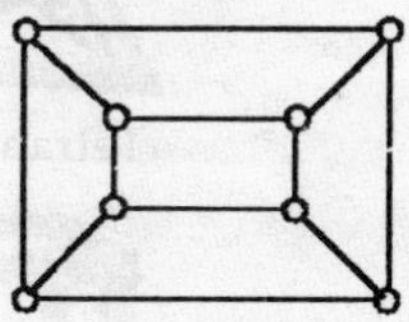

Dodecahedron

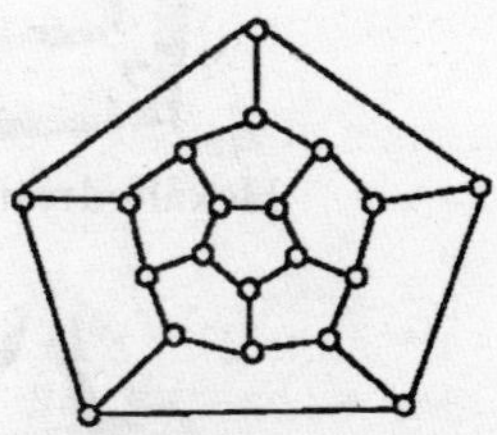

Isosahedron

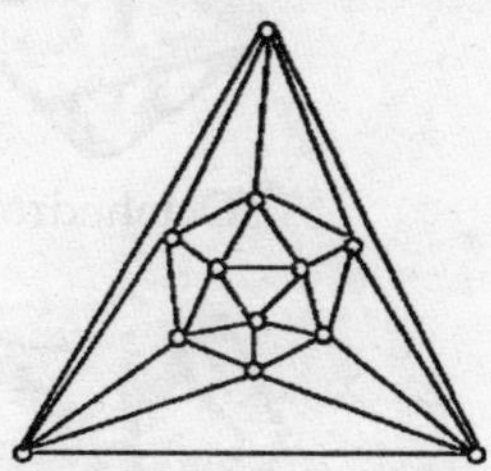

EULER GRAPHS

Consider the following road map

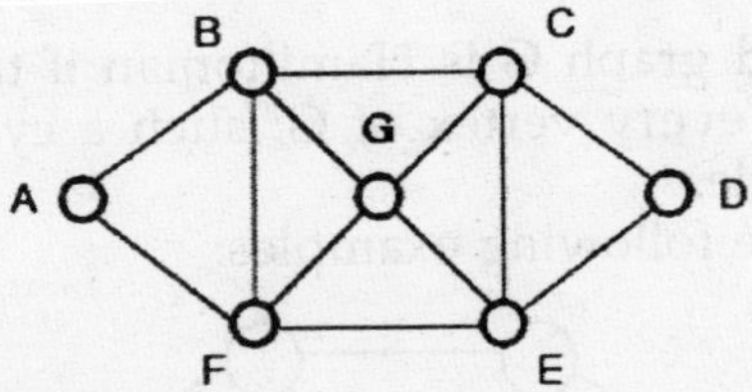

The explorer's Problem: An explorer wants to explore all the routes between a number of cities. Can a tour be found which traverses each route only once? Particularly, find a tour which starts at A, goes along each road exactly once, and ends back at A.

Examples of such tour are

A	B	C	D	E	F	B	G	C
E	G	F	A					
A	F	G	C	D	E	G	B	C
E	F	B	A					

The Explorer travels along each road (edges) just once but may visit a particular city (vertex) several times.

TRAVELER'S PROBLEM

A traveller wants to visit a number of cities. Can a tour be found which visits each city only once? Particularly, find a tour which starts at A, goes to each city exactly once, and ends back at A.

Examples of such tour are

A	B	C	D	E	G	F	A
A	F	E	D	C	G	B	A

The travelers visits each city (vertex) just once but may omit several of the roads (edges) on the way.

Eulerian Trail

A connected graph G is Eulerian if there is a closed trail which includes every edge of G, such a trail is called an Eulerian trail.

Hamiltonian Cycle

A connected graph G is Hamiltonian if there is a cycle which includes every vertex of G; such a cycle is called a Hamiltonian cycle.

Consider the following examples:

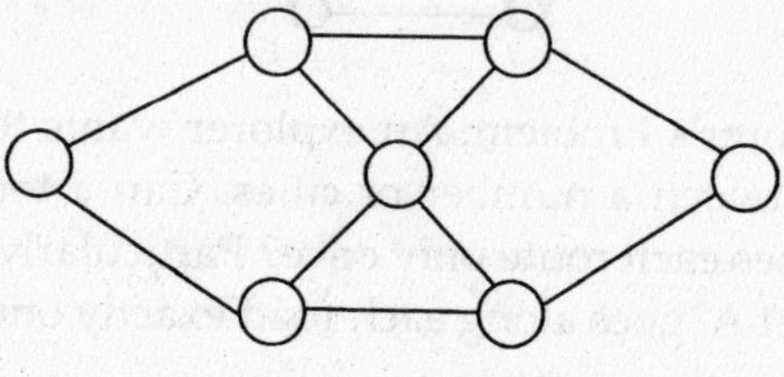

This graph is BOTH Eulerian and Hamiltonian.
This graph is Eulerian, but NOT Hamiltonian.

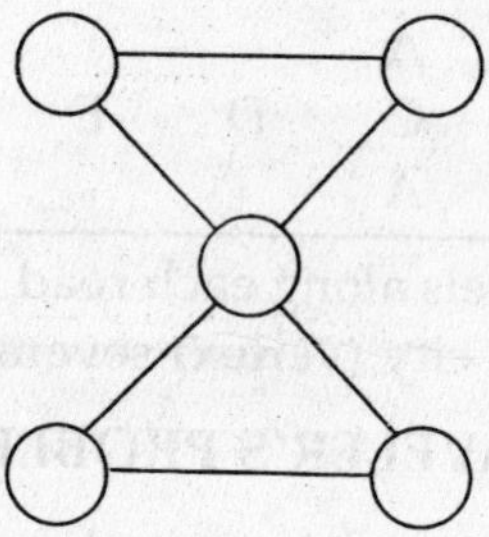

This graph is an Hamiltionian, but NOT Eulerian.

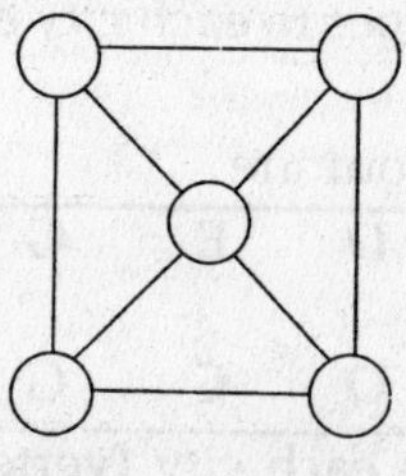

This graph is NEITHER Eulerian NOR Hamiltionian

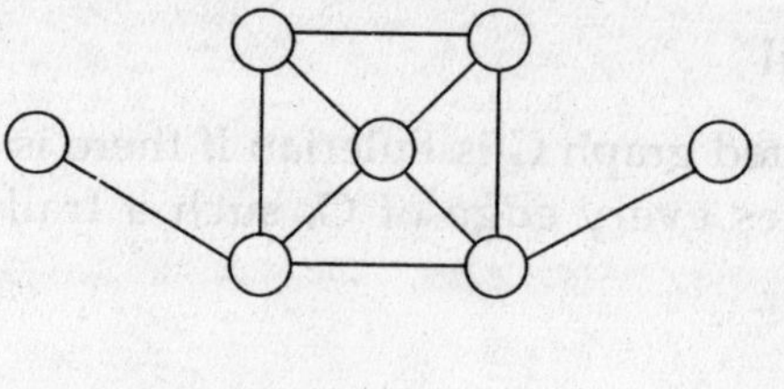

Theorem: Let G be a connected graph. Then G is Eulerian if and only if every vertex of G has even degree.

1. Necessary Condition
   – If G is Eulerian, then every vertex of G has even degree.
2. Sufficient Condition
   – If every vertex of G has even degree, then G is Eulerian.

Eulerian-Type Problems

1. Edge-traceable graphs.
2. Diagrams-Tracing Puzzles.
3. Dominoes.
4. Mazes and labyrinths,
5. The Chinese Postman Problem.
6. The Rotating Drum Problem.

Neither necessary nor sufficient condition is known for a graph to be Hamiltonian. The search for necessary or sufficient conditions is a major area of study in graph theory today.

Sufficient Condition

Dirac's Theorem: Let G be a simple graph with n vertices where n ≥ 3 If deg(v) ≥ 1/2 n for each vertex v, then G is Hamiltonian. For example,

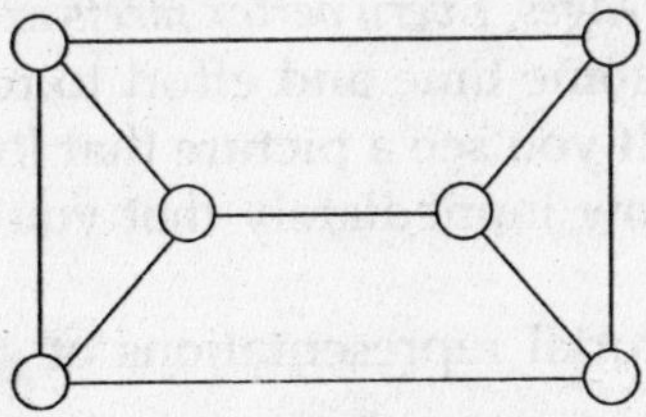

n = 6 and deg(v) = 3 for each vertex, so this graph is Hamiltonian by Dirac's theorem.

Ore's Theorem: *Let G be a simple graph with n vertices where n ≥ 2 if deg(v) + deg(w) ≥ n for each pair of non-adjacent vertices v and w, then G is Hamiltonian.*

For example,

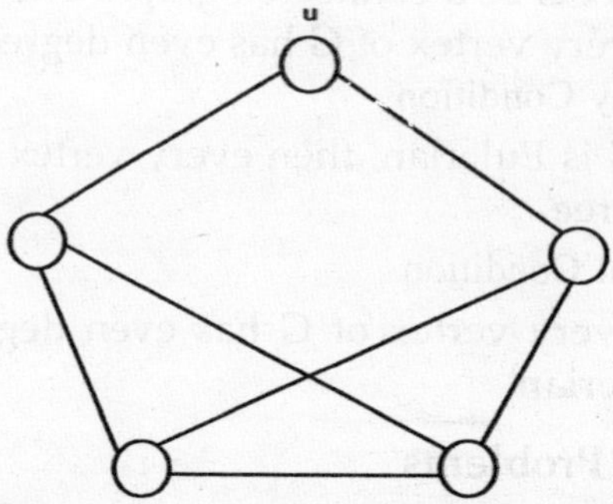

$n = 5$ but $\deg(u) = 2$, so Dirac's theorem does not apply. However, $\deg(v) + \deg(w) \geq 5$ for all pairs of vertices v and w (infact, for all pairs of vertices v and w), so this graph is Hamiltonian by Ore's theorem.

Note that if $\deg(v) \geq 1/2\, n$ for each vertex, then $\deg(v) + \deg(w) \geq n$ for each pair of vertices v and w. It follows that Dirac's theorem can be deduced from Ore's theorem, so we prove only Ore's threoem.

GRAPHS FROM VIEWPOINT

Pictures of graphs showing all of the edges and vertices are great ways to represent a graph. It is immediately clear which vertices are joined by edges, and you can often recognize the graph from its picture. For example, if you were given a very abstract description of the cube graph like: *The graph has eight vertices and 12 edges. Every vertex meets exactly three vertices.* It might take you some time and effort to recognize that this *is* the cube graph. If you see a picture that looks exactly like a cube, then you know immediately that you are dealing with the cube graph.

However, pictorial representations of graphs have their limitations, too.

For example, if you wanted to prove a theorem about graphs, and all you had were pictorial representations, then your proof might have to consist of drawing every single, possible graph, and then demonstrating the truth of your theorem for each picture that you have drawn. A pretty time-consuming process, to say the least.

To help us find other ways of describing graphs, consider the following definitions:

Definition: A graph G consists of a set of vertices and a set of edges, where each edge joins an unordered pair of vertices. The set of vertices of G is denoted by $V(G)$ and the set of edges is denoted by $E(G)$.

Definition: A loop is an edge that joins a vertex to itself. A multiple edge occurs when there is more than one edge joining a pair of vertices.

Definition: A simple graph is a pair $(V(G), E(G))$ where $V(G)$ is a non-empty finite set of elements, and $E(G)$ is a finite set of unordered pairs of distinct elements of $V(G)$.

ABSTRACT DEFINITIONS AND EXAMPLES OF GRAPHS

Pictorial representations of graphs are really useful representations, but are sometimes inconvenient to use when proving theorems about combinatorial graphs. For example, planar graphs (graphs that can be drawn on a page so that no edges cross) obey the following formula:

$$v - e + f = 2$$

wheıe v represents the number of vertices of the graph, e is the number of edges that the graph has, and f is the number of faces that the graph has.

1. How could you prove this formula?
2. If you took the summer course *Geometry and the Imagination*, then you might remember that we proved a similar formula:

$$v - e + f = 2k$$

where $k = 1$ for a sphere, $k = 0$ for a Euclidean plane, and $0k = 1$ for a hyperbolic plane. We did the case of the sphere in class. Remember that we put a 'grid' of triangles on the sphere, and then counted up the number of faces, edges and vertices. The formula we got for the *sphere* was:

$$v - e + f = 2$$

which is exactly the same as the formula for the *planar* graph (i.e. graph that you can draw in the *Euclidean plane,* a.k.a. a piece of paper). Why is the formula for a *planar* graph the same as the geometry formula for the *sphere* (and not the same as the geometry formula for the *Euclidean plane*)?

PICTURE OF GRAPHS VERSUS ABSTRACT DEFINITIONS

One of the approaches that you might have tried in the thinking opportunity above is to draw some examples of planar graphs, and count the numbers of edges, vertices and faces to check to make sure that the formula is correct. Here are some examples:

THE TETRAHEDRAL GRAPH

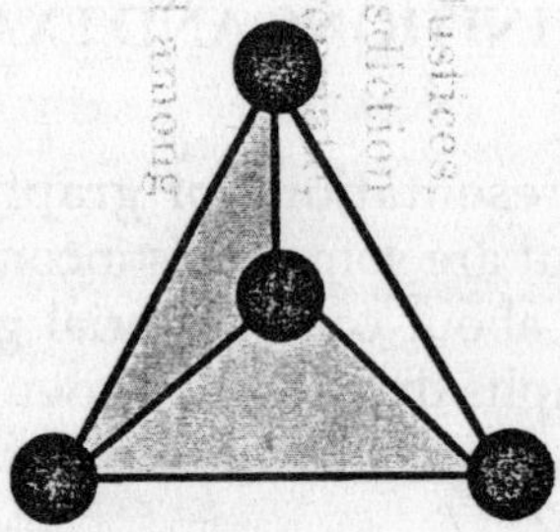

There are 4 *vertices*. $v = 4$.
There are 6 *edges*. $e = 6$.
There are 4 *faces*. $f = 4$.
So: $v - e + f = 4 - 6 + 4 = 2$.

THE CUBE GRAPH

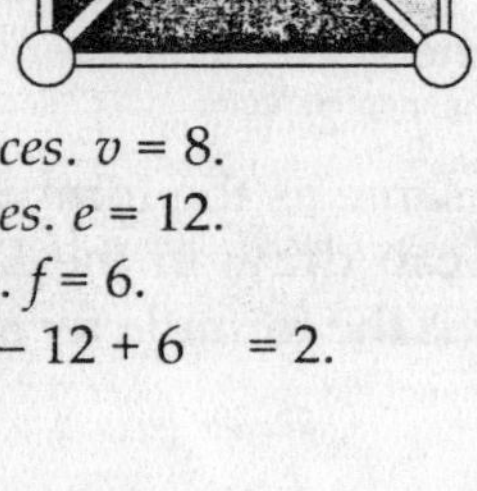

There are 8 *vertices*. $v = 8$.
There are 12 *edges*. $e = 12$.
There are 6 *faces*. $f = 6$.
So: $v - e + f = 8 - 12 + 6 = 2$.

THE DODECAHEDRAL GRAPH

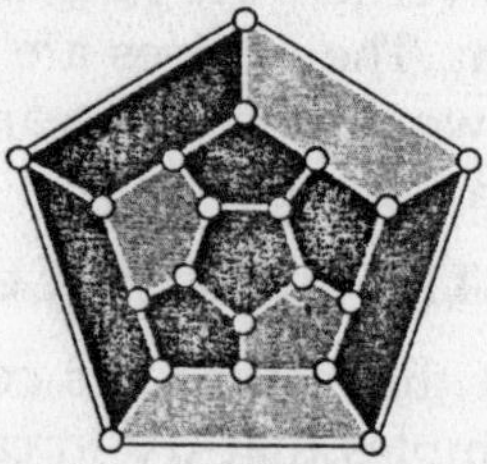

There are 20 *vertices*. $v = 20$.

There are 30 *edges*. $e = 30$.

There are 12 *faces*. $f = 12$.

So: $v - e + f = 20 - 30 + 12 = 2$.

These examples verify the formula, and the pictorial representations of the graphs make it very easy to identify and count the numbers of edges, vertices and faces for each graph. However, if we were to try and prove the formula:

$$v - e + f = 2$$

using only these pictorial representations of the graphs, then we would have to draw out *every* single planar graph, and verify the formula for each picture that we drew. Since there are a *lot* of possible planar graphs, this would be an impractical approach. One answer to this is to make *abstract* definitions. These definitions are useful because they let us prove things more easily, but they are usually harder to understand than the pictorial representations of graphs.

Abstract Definitions

Definition: A graph G consists of a set of vertices and a set of edges, where each edge joins an unordered pair of vertices. The set of vertices of G is denoted by $V(G)$ and the set of edges is denoted by $E(G)$.

Definition: A loop is an edge that joins a vertex to itself. A multiple edge occurs when there is more than one edge joining a pair of vertices.

Definition: A simple graph is a pair $(V(G), E(G))$ where $V(G)$ is a non-empty finite set of elements, and $E(G)$ is a finite set of unordered pairs of distinct elements of $V(G)$.

New for this lesson:

Definition: Two vertices are said to be adjacent if there is an edge joining them. The vertices are said to be incident to the edge. Similarly, two distinct edges are adjacent if they have at least one vertex in common.

PATHS AND CIRCUITS

So far, much of this course has consisted of advancing definitions and examples that try to capture the meaning if mathematical concepts in abstract language. Let's try to see how some of these definitions might be used to decide theoretical and practical questions. If you are interested in the theoretical side of mathematics, then you might like to consider the following proposition.

Proposition: Suppose that G is a graph, with $n > 0$ vertices. Let $d_1, \ldots, d_n$ denote the degrees of the vertices of the graph G (see lesson 2 for the definition of the degree of a vertex). Then the sum of the degrees is an even number, and can be written in the form:

$$\sum_{j=1}^{n} d_j = 2m$$

1. This proposition is *not true*. There are graphs that satisfy the hypotheses of the proposition but do not obey the conclusion. See if you can find an example of a graph that does not satisfy the proposition.
2. The proposition can be fixed to be mathematically true by adding hypotheses that the graph must satisfy. Can you find hypotheses that make the proposition true? You might like to consider the example that you came up with, and try to identify what features of your example mess up the proposition.
3. The number $m >= 0$ is not the same for every graph. You might like to try some examples of graphs that satisfy your hypotheses to get some ideas for what m might be.
4. See if you can use the definitions and concepts outlined in the first two lessons to develop a mathematical proof of the proposition (the version that includes your hypotheses).

If you are interested in applying mathematics to solve

practical problems, you might like to think about the following statement: *In a gathering of six people, either there are three people who all know each other, or else there are three people none of whom knows the other two.*

1. How could you use a graph to represent the gathering of six people, and how could you represent who knows who.
2. The Latin phrase, "Nosce te ipsum," can be translated as, "Know thyself." How would you represent the possibility of people at the gathering 'knowing themselves' on your graph? If people at the gathering are allowed to 'know themselves,' is the statement about who knows who at the gathering still valid?
3. If you assume that there are three people at the gathering who all know each other, how would you represent this using a graph? Let's call a graph like this a 'Type A' graph.
4. If, instead, you assume that there are three people at the gathering, none of whom knows the other two, how would you represent this using a graph? Let's call a graph like this a 'Type B' graph.
5. How could you show that any possible graph that represents the gathering of six people has to be a 'Type A' of a 'Type B' graph?
6. You might like to consider the following line of reasoning, and try to apply it to the questions posed in this thinking opportunity.

We could represent the people at the gathering by vertices on a graph. A gathering of six people would be represented by a graph with six vertices. If two people at the gathering know each other, then we could draw an edge between the two vertices. (Here, the possibility of people knowing themselves is ignored.) For example, a gathering might resemble:

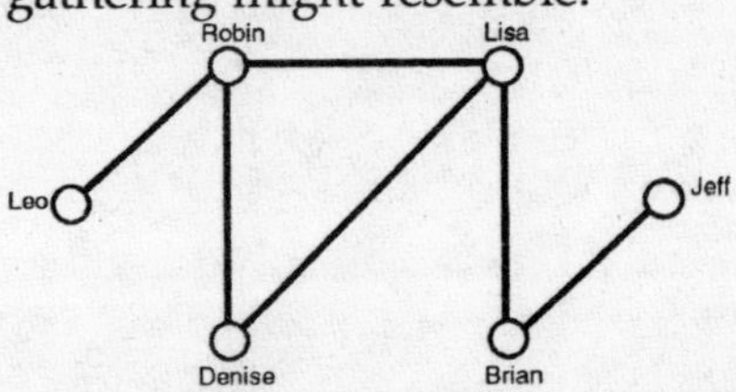

At this gathering, Robin, Lisa and Denise all know each

other. Graphically, the edges linking Robin's, Lisa's and Denise's vertices forms a triangle.

Another gathering might look like:

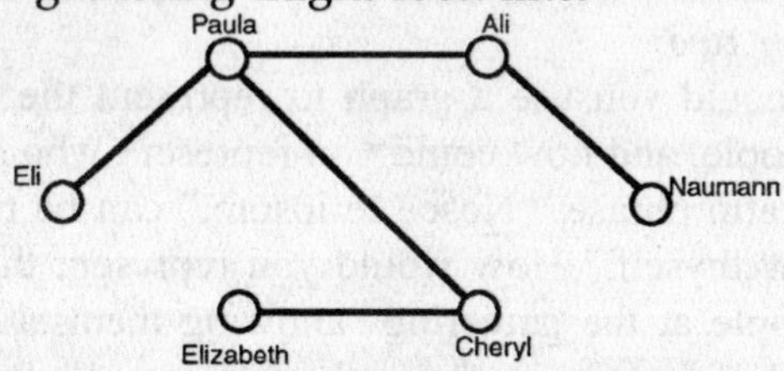

Here, Ali, Eli and Elizabeth do not know each other. There is a graph, based on the pink and green graph, that has an edge between two vertices if the two vertices *don't* have an edge between them in the pink and green graph. This is called the *conjugate* of the pink and green graph, and resembles:

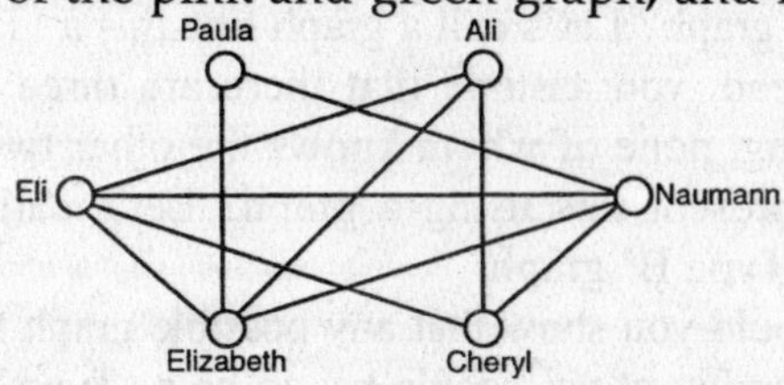

Chapter 7

Tree

INTRODUCTION

In graph theory, a tree is a graph in which any two vertices are connected by exactly one path. Alternatively, any connected graph with no cycles is a tree. A forest is a disjoint union of trees. Trees are widely used in computer science data structures such as binary search trees, heaps, tries, Huffman trees for data compression, etc. In some of the previous lessons, there have been thinking opportunities that relate mathematically defined words (such as path and circuit) to the everyday meaning of the words.

A tree is an undirected simple graph G that satisfies any of the following equivalent conditions:

1. G is connected and has no cycles.
2. G has no cycles, and a simple cycle is formed if any edge is added to G.
3. G is connected, and it is not connected anymore if any edge is removed from G.
4. G is connected and the 3-vertex complete graph K_3 is not a minor of G.
5. Any two vertices in G can be connected by a unique simple path.

If G has finitely many vertices, say n of them, then the above statements are also equivalent to any of the following conditions:

1. G is connected and has $n - 1$ edges.
2. G has no simple cycles and has $n - 1$ edges.

An undirected simple graph G is called a *forest* if it has no simple cycles.

A directed tree is a directed graph which would be a tree if the directions on the edges were ignored. Some authors restrict the phrase to the case where the edges are all directed towards a particular vertex, or all directed away from a particular vertex.

A tree is called a rooted tree if one vertex has been designated the *root*, in which case the edges have a natural orientation, *towards* or *away* from the root. The *tree-order* is the partial ordering on the vertices of a tree with *u* d" *v* if and only if the unique path from the root to v passes through u. A tree which is a subgraph of some graph G is a normal tree if the ends of every edge in G are comparable in this tree-order. Rooted trees, often with additional structure such as ordering of the neighbors at each vertex, are a key data structure in computer science; see tree data structure. In a context where trees are supposed to have a root, a tree without any designated root is called a free tree.

A polytree has at most one undirected path between any two vertices. In other words, a polytree is a directed acyclic graph (DAG) for which there are no undirected cycles either.

A labeled tree is a tree in which each vertex is given a unique label. The vertices of a labeled tree on n vertices are typically given the labels 1, 2, ..., n. A recursive tree is a labeled rooted tree where the vertex labels respect the tree order (*i.e.*, if $u < v$ for two vertices u and v, then the label of u is smaller than the label of v). An irreducible (or *series-reduced*) tree is a tree in which there is no vertex of degree 2. An ordered tree is a tree for which an ordering is specified for the children of each node. An n-ary tree is a tree for which each node which is not a leaf has at most n children. *2-ary trees* (resp. *3-ary trees*) are sometimes called binary trees (resp. ternary trees) For example, a mathematical path might resemble:

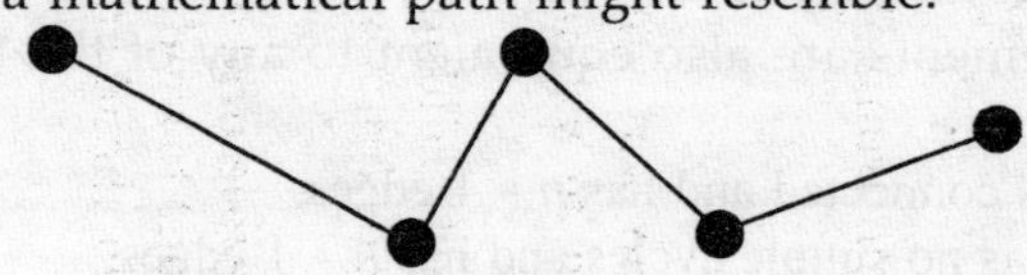

The everyday use of the word path might evoke an image resembling:

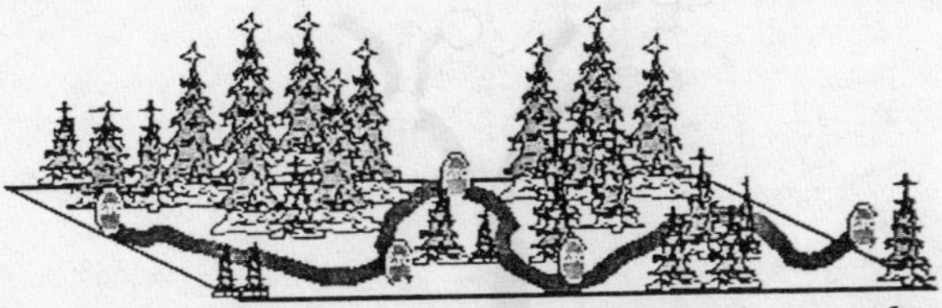

There is a correspondence between the mathematical meaning of *path*, and the image evoked by the everyday sense of the word *path*. In this instance, the vertices are given by the trail marker sings, and the edges are given by the sections of the trail joining the trail markers. As you might expect, it is possible to see associations between the mathematical meaning of the word *tree* and the images evoked by the everyday meaning of the word *tree*. A *tree* is basically a graph that can be drawn to look something like:

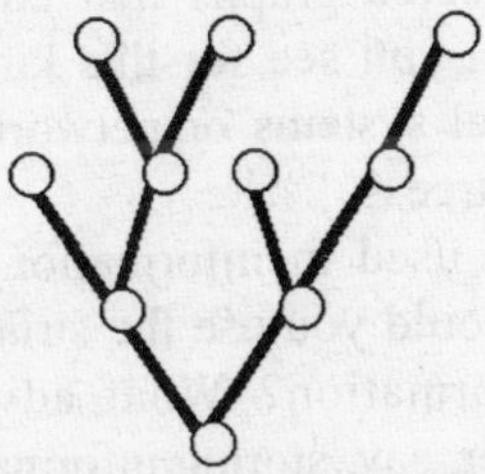

In my mind, the everyday meaning of the word *tree* evokes an image like:

Drawing vertices at the ends of the branches, and at the places where branches split gives a picture like:

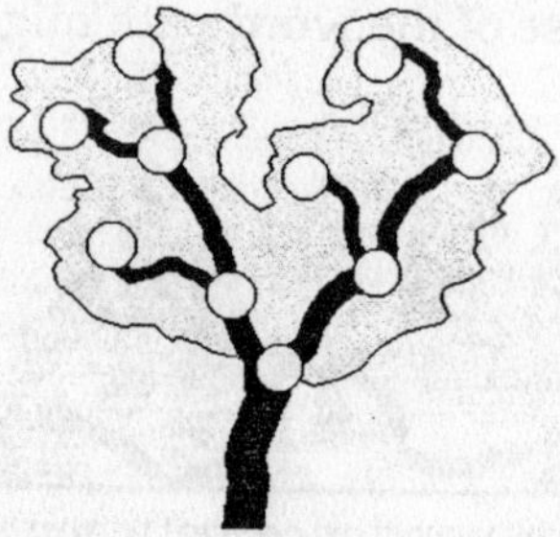

The branches of the (everyday meaning) *tree* can then be used for the edges of the (mathematical meaning) *tree* (the trunk of the tree doesn't connect two vertices, so it can't be used as an edge).

So, the mathematical meaning of the word *tree* can be related quite directly to the 'everyday' meaning of the word *tree*. However, this does not (at least in my mind) suggest what (mathematical) trees might be useful for.

1. Trees are connected graphs that contain no loops. What applications can you see for this kind of graph? Can you think of physical systems or networks that would be well represented by trees?
2. Trees are often used in information storage and retrieval systems. How could you use the structure of a tree to store or catalog information? What advantages would your system have over, say, storing information sequentially (that is, like a bunch of 35 cards in a card file box)?

AN APPLICATION FOR TREES: SORTING ALGORITHMS

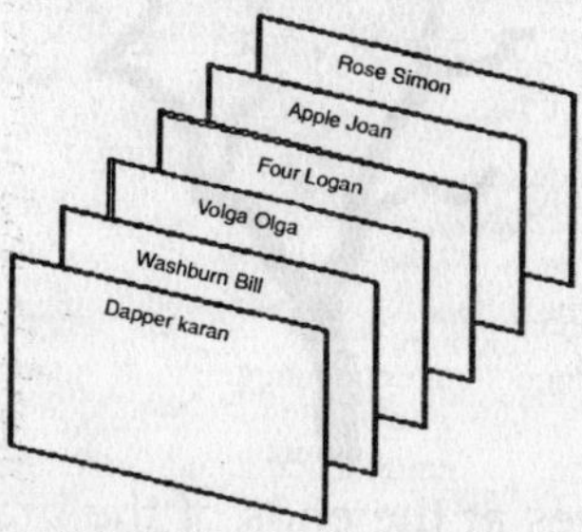

A sorting algorithm is an outline or description of a

procedure for arranging data in a particular order. For example, you might (like me!) have trouble remembering important information like people's birthdays, and keep a card file.

This file isn't very organized at the moment - you might end up having to go through every person's card to find the one you are looking for. To me, the most immediate way of organizing this file would be to arrange it in alphabetical order, by last name.

For a file of only six cards, you could just pick them all up, and put them back in the file box in alphabetical order. If your file had thousands of cards, then you would have to be more systematic. For example, you could go through the cards, comparing each card with the one immediately behind it. If the cards are out of alphabetical order, then you would switch them.

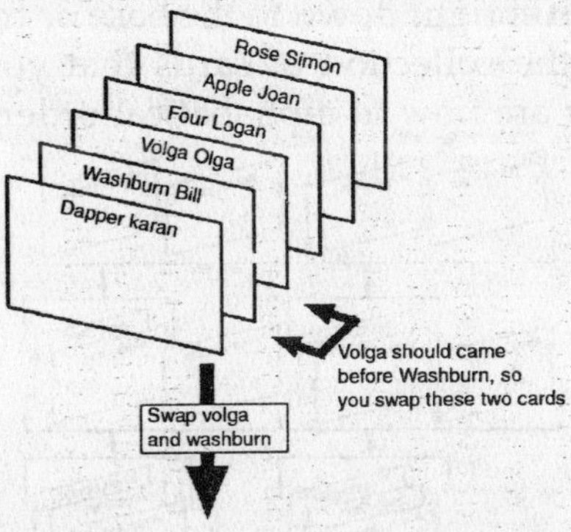

This kind of sorting algorithm, where you compare the cards pair by pair and swap them if the pair is not in alphabetical order is often called a *bubble sort*. As you can probably anticipate, bubble sorting is a tedious and time-consuming process (even for a computer).

Trees can be used to sort the cards into alphabetical order in fewer steps. The idea behind using trees to sort a list of cards is to split the collection of cards into a lot of smaller collections, and then to split these into even smaller collections. For example, with the file of cards used to illustrate the bubble sort, the collection can be split up as shown below.

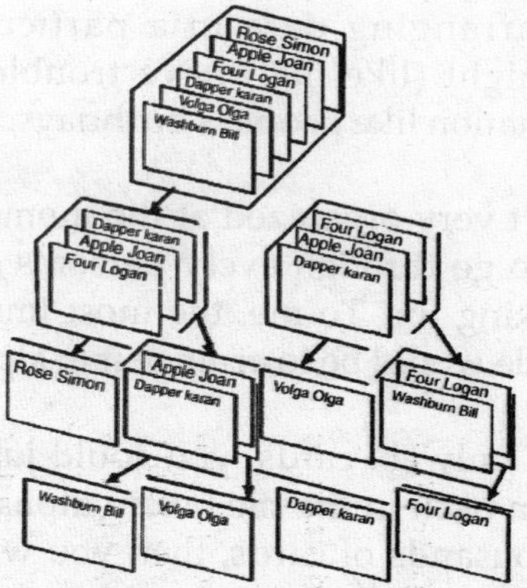

To sort the cards, the tree is turned upside down, and the cards recombined into steadily larger and larger collections. When collections of cards are put together, they are merged so that the resulting collection is in alphabetical order. When the process has run right down to the base or trunk of the tree, then you have the collection of cards that you started with, except that they are now in alphabetical order.

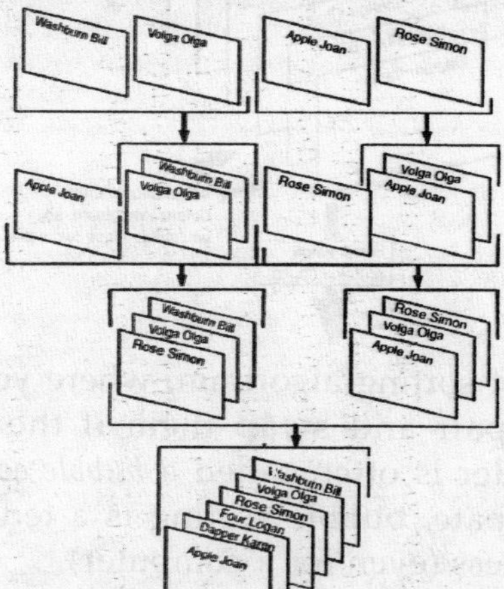

The mathematical definitions of trees and forests are given below. See if you can see how these definitions and the everyday notions of *tree* and *forest* correspond to each other.

Definition: A forest is defined to be a graph that contains no circuits. A tree is a connected forest.

Theorem: Suppose that T is a graph with n vertices. The following statements are equivalent:

1. T is a tree.
2. T has $n-1$ edges and no circuits.
3. T is connected and has $n-1$ edges.
4. T is connected and each of its edges is an isthmus.
5. Any two vertices of T are connected by exactly one simple path.
6. T has no circuits, but the addition of any new edge creates exactly one circuit.

Note: The theorem states that all of the statements are equivalent. Two statements, *A* and *B*, are equivalent provided that statement *A* is true if and only if *B* is true. This is another one of those "double barreled" statements, where you have to prove "both barrels" in order to prove the theorem. In this case, we would have to show that if statement 1 is true, then statement 2 is true. Then, to do the other "barrel" of the proof, we'd have to show that if statement 2 is true, then statement 1 is true. This would have to be repeated for every pair of distinct statements in the theorem, and would amount to 30 little proofs that together constitute the proof of the theorem. This would be a pretty big job!

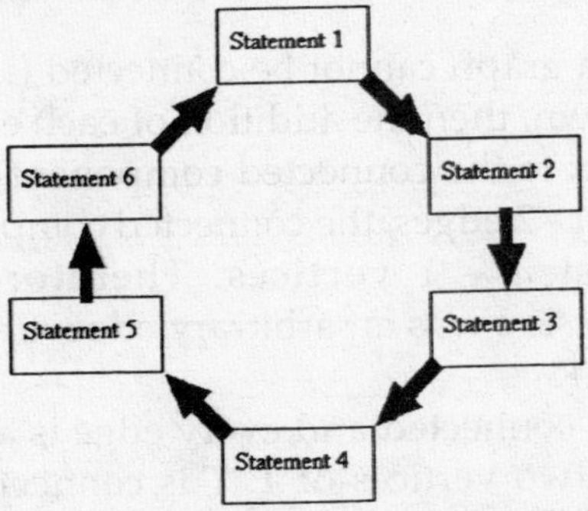

With a little organization, the proof of the theorem can be cut down to just proving six little proofs; that if statement 1 is true then statement 2 is true, if statement 2 is true then statement 3 is true, and so on, concluding with the proof that if statement 6 is true, then statement 1 is true. The proof ends up as a cycle of logical inferences.

Proof: If $n = 1$, all six results are trivially true. So, assume $n >= 2$.

Use induction. Suppose the result is true for all graphs with fewer with n vertices.

Remove any edge, say from v to w. In the resulting graph, v and w lie in distinct connected components, A and B (each connected component is circled in the above diagram). Each of these connected components is a tree, say on r and s vertices respectively. A has r 1, and B has s *1* edges. Therefore, T has

$$(r - 1) + (s - 1) + 1 = r + s - 1 = n - 1 \text{ edges.}$$

$2 \Rightarrow 3$ T has $n - 1$ edges and no circuits. Assume that:

$$T = C_1 \cup L \cup C_3$$

where C_1 is a connected component of T with n, vertices. There, C, is a tree with n, $-$ 1 edges. So, the number of edges in the graph T is:

$$\sum_{j=1}^{s}(n_j - 1) = \left(\sum_{j=1}^{s} n_j\right) - S = n - S.$$

But, T has $n - 1$ edges, so $s = 1$. Hence, T is connected.

$3 \Rightarrow 4$ T is connected and has $n - 1$ edges. Let e be any edge of T. Remove e and you get a graph with n vertices and $n - 2$ edges.

But such a graph cannot be connected (i.e. start with any vertex on its own, then the addition of each edge in turn adds one new vertex to the connected component containing v. As there are only $n - 2$ edges, the connected component containing v has at most $n - 1$ vertices. Therefore the graph is disconnected). As e was an arbitrary edge, then every edge of T disconnects T.

$4 \Rightarrow 5$ T is connected and every edge is an isthmus. Let v and w be any two vertices of T. T is connected, so there is a simple path from v to w. Assume that there are 2 or more such paths. Then the removal of any edge of the second path that is not part of the first path does not disconnect the graph, T. Therefore, that edges removed was not an isthmus, giving a contradiction. Therefore, there is only one simple path from v to w in T.

5 ⇒ 6 Every two vertices are connected by a simple path. If T contained a circuit, then any two vertices in this circuit must be connected by at least two simple paths, contradicting 5, implying that T can contain no circuit. Add a new edge, say from x to y. This creates a circuit because x and y are already connected in T. If more than one circuit was created, then x and y would have to be joined by two or more paths in T, which is not the case, so only one circuit is created.

6 ⇒ 1 T has no circuits, but the addition of any edge creates exactly one circuit. Must prove that T is connected. Let v and w be any vertices of T. If no edge exists from v to w then *add* such an edge and you obtain a circuit. If you then remove this edge, then you'll be left with a path from x to y in T, so T is connected. This completes the proof of the theorem.

SUBTREES AND SPANNING TREES

Definition: A subtree of a graph G is any subgraph of G that is a tree. A sub-forest is any subgraph of G that is a forest.

Definition: If G is a connected graph, then a spanning tree for G is any subgraph of G that is a tree, and includes *every* vertex of G.

Example: One pictorial representation of the cube graph is:

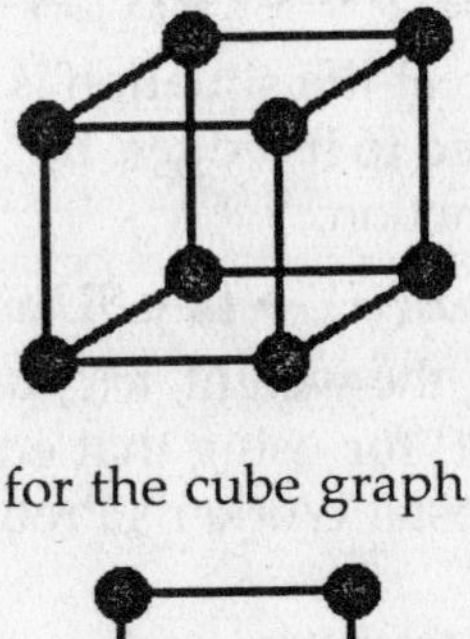

A spanning tree for the cube graph is:

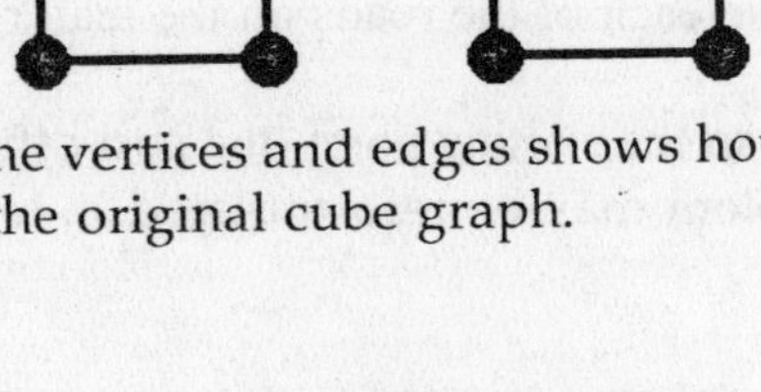

Coloring the vertices and edges shows how the spanning tree relates to the original cube graph.

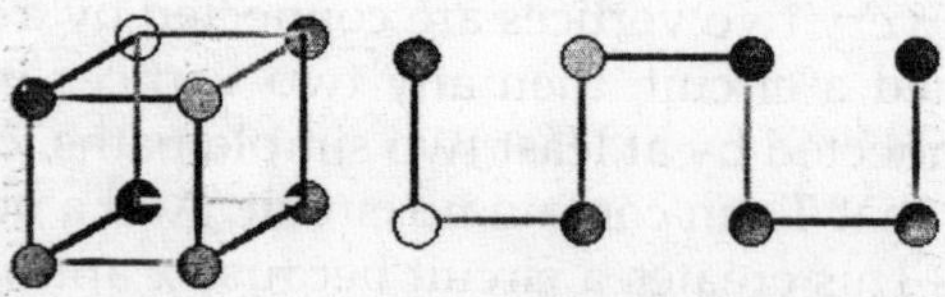

Proposition: For $n >= 2$, the complete graph on n vertices (i.e. K_n) has n^{n-2} different spanning trees.

Hint: You might approach this proof by examing the first few cases of the proposition ($n = 2$, $n = 3$, etc.) and then try to use the technique of proof by induction.

Definition: Any vertex, v, of a connected graph G with the property that the maximum distance from v to any other vertex is as small as possible is called a *central vertex* (or a centre of the graph G). That maximum distance is called the *radius* of G.

Example: The graph,

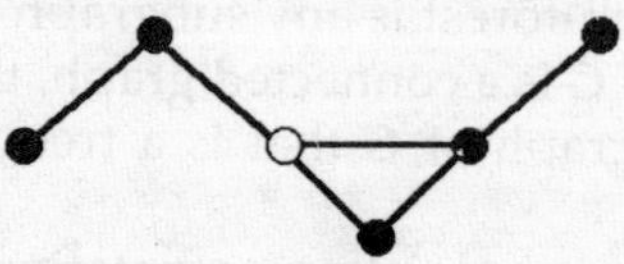

has radius 2 (the central vertex is shaded yellow).

THE MINIMAL CONNECTOR PROBLEM

Suppose some real-life situation is modeled by a graph G with weights assigned to its edges. By 'weights' we mean that there is a weight function,

$$w : E(G) \rightarrow \{x \in \Re : x \geq 0\},$$

and for each edge, e, the weight, $w(e)$ usually represents some sort of cost (or profit) for using that edge.

Example: The Postal Worker's Problem. The postal worker wants to:

1. Deliver all of the letters, and,
2. Cover the least possible distance when doing so, and,
3. Traverse each of the roads on the mail route at least once, and,
4. Return to the starting point (the post office).

This problem may be reformulated in terms of weighted

graphs, where the graph corresponds to the network of roads, and the weight of each edge corresponds to the length of the road represented by that edge.

The requirement is to find a closed walk that includes every edge at least once, and has the least possible weight.

Example: The Traveling Sales Representative's Problem. The sales representative

1. Has to visit several different cities, then return to the starting point, and,
2. Travel the least possible distance.

This can also be reformulated in terms of weighted graphs. In this case, the requirement is to find a Hamiltonian cycle of least possible weight in a complete graph with weights.

Often, you want to find a connected subgraph of a graph G that uses every vertex of G (e.g. so you can go from any vertex to any other vertex in the subgraph) but so that the sum of the weights of all the edges in the subgraph is as small as possible. This is called the Minimal Connector Problem. The solutions to this problem is given by Kruskal's Algorithm.

This is an algorithm for finding a minimum weight spanning tree in a connected graph G that has weights assigned to its edges - in particular, it solves the minimal connector problem for G.

Definition: If T is any subgraph, define $w(T)$ = sum of the weights of the edges of T. A *minimum weight spanning tree* is a spanning tree for which $w(T)$ is as small as possible.

KRUSKAL'S ALGORITHM

First choose any edge e_1 of smallest weight in G. Then choose edges e_2, e_3,... by choosing at each stage an edge (of smallest weight not already chosen) subject to the condition that it forms no circuit with any of the previous edges. If there are no such edges left, then what you have already chosen are the edges of a minimum weight spanning tree.

Example: A cube graph with weights, and a minimum spanning tree. The weight of the minimum weight spanning tree is 34.

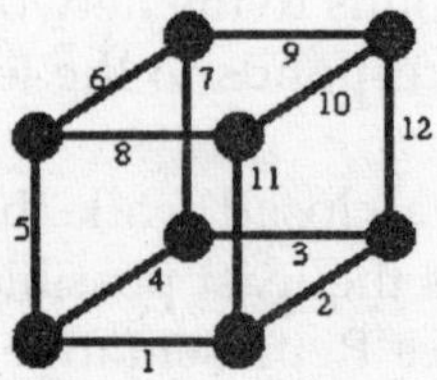

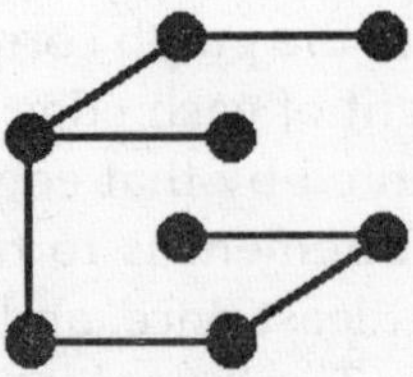

Proof: (That Kruskal's Algorithm always works for a connected graph G.)

First, note that if $e_1, e_2, \ldots, e_m$ are the edges chosen by the time the algorithm finishes, then these edges must form a *connected* subgraph of G, that uses every vertex of G. (If not, then you would have a forest, not a tree, but G is connected so there will be some edges that could not be some edges that could be chosen as e_{m+1}, etc.)

It is necessary to prove that T is a minimum weight spanning tree. Let S be any spanning tree for G (of minimum weight). Consider e_1 to e_m in turn. Choose the first one of these that *isn't* an edge of S, say e_k. Add e_k to S. This creates a circuit in S that has an edge e that can't be in T (as T has no circuits). By the choice of e_k, $w(e_k) =< w(e)$. Remove e from S and replace it by e_k. This gives a new spanning tree S' with $w(S') =< w(S)$. Continue in this way, choosing the next e_l that isn't in S', etc. Eventually you get:

$$w(T) \leq \mathbb{K} \leq w(S''') \leq w(S'') \leq w(S') \leq w(S) \leq w(T).$$

So, $w(T) = w(S)$. Since S is a minimal spanning tree, T is a minimal spanning tree, too. This completes the proof.

PLANAR GRAPHS

Definition: A *planar graph* is one that can be drawn on a plane in such a way that there are no "edge crossings," i.e. edges intersect only at their common vertices.

Example: A pictorial representation of the cube graph that makes it easy to see why it is called the cube graph resembles:

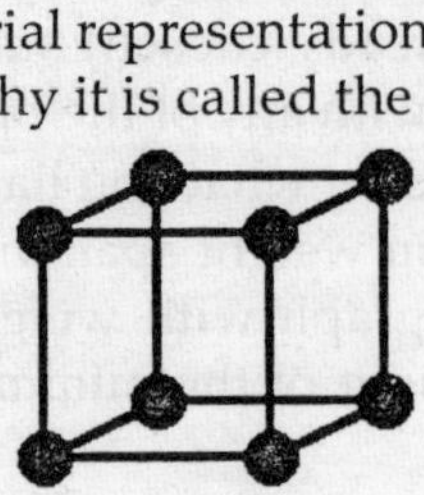

This representation includes many "edge crossing." It is possible to re- draw the cube graph so that no two edges cross. Coloring the vertices and edges shows how the graph is re-drawn morc clearly.

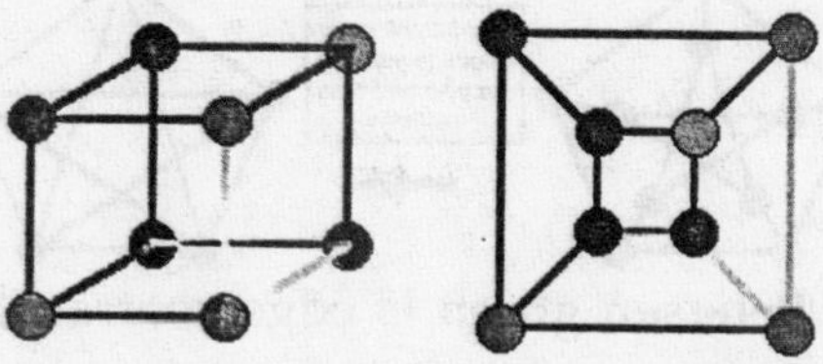

Definition: A *plane graph* is one that has been drawn in the plane in such a way that its edges intersect only at their common end-vertices.

Definition: Two graphs, *G* and *H* are defined to be homeomorphic if you can make one graph into the other by inserting vertices of degree 2

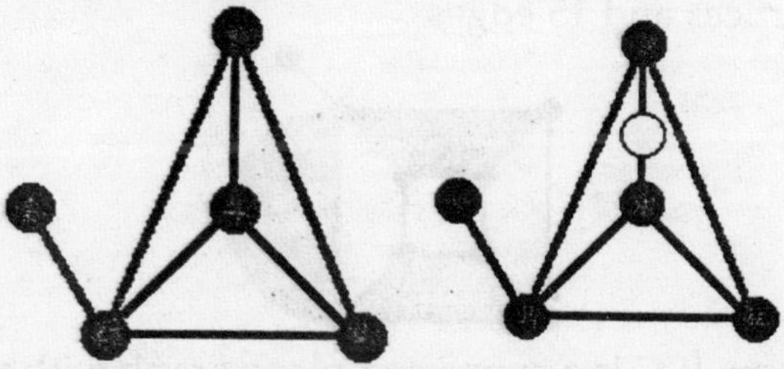

Theorem: (Kuratowski's Theorem) A finite graph *G* is planar if and only if it has no subgraph that is homeomorphic to the complete graph in five vertices (K_5) or the complete bipartite graph $K_{3,3}$.

Definition: Examples of edge contractions.

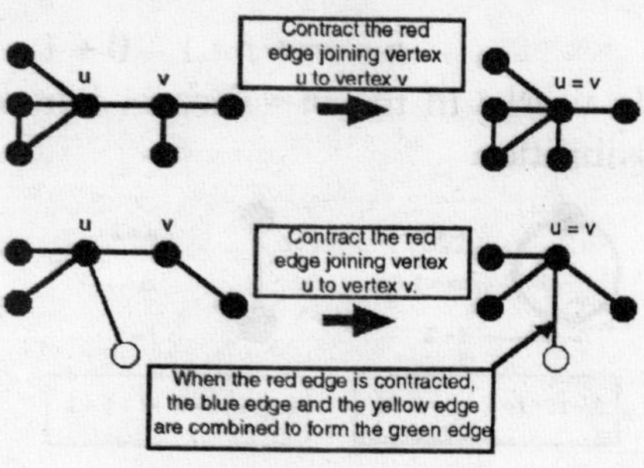

Theorem: A finite graph G is planar if and only if it contains no subgraph that is edge- contractible to K_5 or $K_{3,3}$.

Example: The Petersen graph may be represented pictorially as:

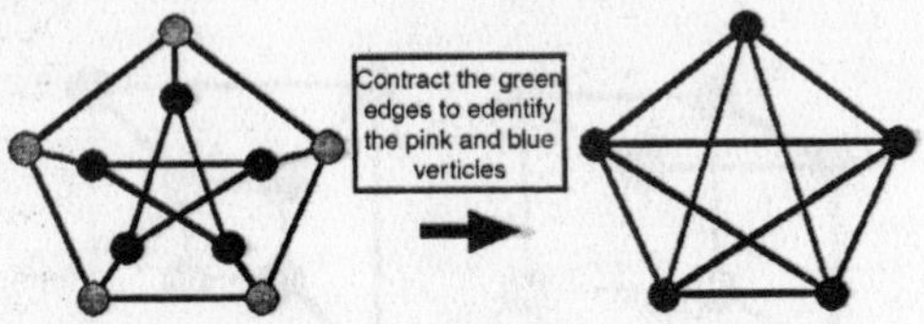

Since the Petersen graph is edge contractible to K_5, it is not a planar graph.

EULER'S POLYHEDRON FORMULA

Definition Let G be a plane graph, and consider the regions bounded by the edges of G. These are called *faces*.

Example The following graph has 8 faces (one of them is colored - the outside of the graph is usually called the infinite face), 9 vertices and 15 edges.

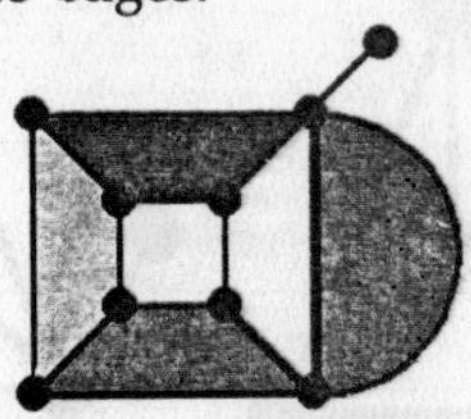

Theorem: If G is a connected plane graph with n vertices, m edges and f faces, then:

$$n - m + f = 2.$$

Proof: (By induction on m.) Suppose that $m = 0$. Then for G to be a connected graph, there may be only one vertex, and one face. So

$$n - m + f = 1 - 0 + 1 = 2.$$

and the formula works in the $m = 0$ case. For the $m = 1$ case, there are 2 possibilities.

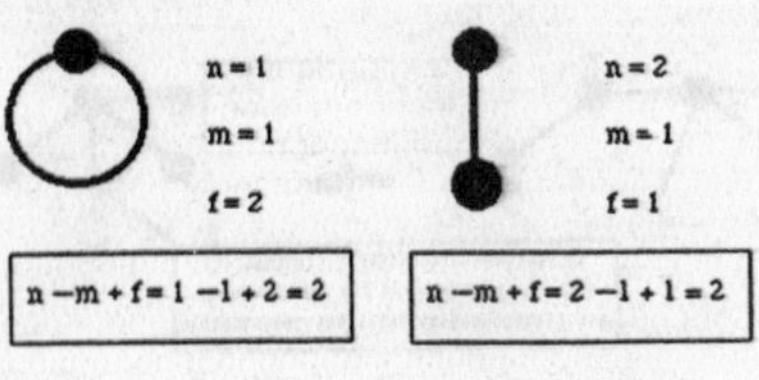

So, Euler's Polyhedron Formula works for $m =< 0$. For the inductive step, assume that $m > 1$ is given, and that the Euler Polyhedron Formula works for all graphs with m - 1 edges.

If G is a tree, then $f = 1$ and $m = n$- 1. So,

$$n - m + f = n - (n - 1) + f = n - n + 1 + 1 = 2.$$

provided that G is a tree. If G is not a tree, then remove an edge e belonging to some circuit. After removing the edge, e, you are left with a plane graph H on n vertices with $m - 1$ edges. Denote the number of vertices of H by n_H, the number of edges of H by m_H, and the number of faces of H by f_H. Then, $n_H = n$, $m_H = m - 1$, and $f_H = f - 1$. Since $m_H = m - 1$, H satisfies the Euler Polyhedron Formula, so:

$$2 = n_H - m_H + f_H = n - (m - 1) + (f - 1) = n - m + f + 1 - 1 = n - m + f.$$

Thus, G satisfies the Euler Polyhedron Formula. This completes the proof of the theorem.

Corollary: If G is a (connected) simple, finite planar graph with n vertices ($n > = 3$), then G has at most $3n$ - 6 edges. Furthermore, if G contains no triangles, then G has at most $2n - 4$ edges.

Proof: Suppose that G has at most m edges, consider some plane drawing of G, with f faces. Consider the number of pairs (e, F) where e is one of the edges bounding the face F, viz:

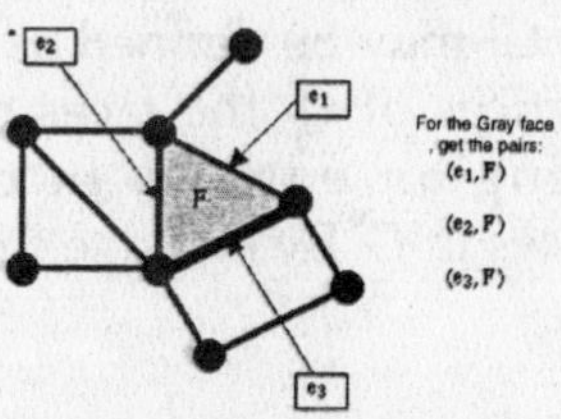

For each edge e, there are at most 2 faces that it bounds. So, the total number of these edge face pairs has to be less than $2m$. On the other hand, because G is a simple graph, each face is bounded by at least 3 edges. Therefore, the total number of edge-face pairs is greater than or equal to $3f$.

So, $3f = < 2m$.

By the Euler Polyhedron Formula, $n - m + f = 2$,, so, $3n - 3m + 3f = 6$. Since $3f = < 2m$,

$$3f = 6 - 3n + 3m \leq 2m$$

Re-arranging,

$$m \leq 3n - 6.$$

Next, if G has no triangles, each face must be be bounded by 4 or more edges. Then $2f = < m$, and $m = < 2n - 4$. This completes the proof of the corollary.

Note: This corollary is true even for disconnected, simple, finite planar graphs, provided that each connected component has at least three vertices.

Corollary: Every finite, simple, planar graph has a vertex of degree less than 6.

Proof: Suppose that G is such a graph. Without loss of generality, G is connected. Suppose that G has n vertices, m edges, and that vertex j has degree d_j (for $1 = < j = < n$). Then,

$$\sum_{j=1}^{n} d_j = 2m \leq 2(3n - 6) = 6n - 12$$

Then, the average degree of the vertices of the graph G, is given by,

$$\textit{Average Degree} = < \frac{6n - 12}{n} = 6 \frac{12}{n} < 6..$$

So, at least one of the vertices has degree less than 6. This completes the proof of the corollary.

Definition: Suppose that G is a (finite) connected plane graph, then its dual G^* may be defined as follows:

Take as the vertices of G^* the faces of G, and for every pair of faces having an edge, e, in common, join the corresponding vertices in G^* by an edge that crosses the edge, e, only once.

Chapter 8

Lead-Lag Algorithms

OVERVIEW

The derivations of recurrence formulas used in the digital simulation of continuous dynamic response, as modeled by ordinary linear differential equations. Emphasis is placed on first and second order equations, since these are the two most common "transfer functions" used in real-world applications.

Along with the optimum recurrence formulas, a variety of specialized and simplified methods are discussed and evaluated to clarify their underlying assumptions and ranges of applicability. Throughout this document we assume that the time required to perform a recurrence computation is negligible relative to the characteristic times of the dynamic functions being simulated. Also, the "round off" error of the computations is assumed to be negligible. In view of the speed and word size of modern digital processors, both of these assumptions are justified when modeling most large-scale physical phenomena.

DISCRETE-TIME SIMULATION OF FIRST-ORDER

The first order lead/lag transfer function is a commonly used mathematical model for representing simple, continuous, dynamic response. A first order lead/lag coupling of two variables implies the relationship

$$\tau_D \frac{dy}{dt} + y = \tau_N \frac{dx}{dt} + x$$

where x is the independent variable, y is the dependent variable, t is time, and the coefficients t_D and t_N are coupling parameters (not necessarily constant). In the continuous domain this relationship uniquely determines the output response y(t) for any given initial conditions and input function x(t). However, there exists a variety of algorithms for simulating in a digital computer, and these algorithms differ both in their computational complexity and in their computed response characteristics.

There are two reasons for the diversity of digital simulations. First, according to equation the change in the output during any interval of time depends on the value of the input continuously throughout the interval, but in digital systems the input is specified only by a sequence of discrete values. Hence the relationship is underspecified in the discrete-time domain, and can only be solved by making an assumption as to the behaviour of the input variable between the discrete values. Second, practical constraints on execution time, execution frequency, and memory storage often lead to the use of modified solutions based on various simplifying assumptions.

OPTIMUM FIRST ORDER SOLUTION

BASIC ASSUMPTIONS

As noted earlier, equation is indeterminate when the input variable is specified only by a sequence of discrete values, unless some assumption is made as to the behaviour of the input variable between the discrete values. When only two discrete values (the current and one past value) are available at any one time, as is the case in virtually all first-order algorithms, the best possible assumption is that "x" varies linearly with time between the two discrete values. If x were significantly non-linear over the interval, then clearly two values would be insufficient to specify the function. So, we can represent the input function over an interval Δt by

$$x = x_p + \left(\frac{x_c - x_p}{\Delta t} \right)$$

where the subscripts p and c denote "past" and "current"

discrete values, corresponding to the initial and final values for that interval. The derivative of this x(t) during the interval is then just

$$\frac{dx}{dt} = \left(\frac{x_c - x_p}{\Delta t}\right)$$

so equation can be rewritten

$$\tau_D \frac{dy}{dt} + y = \tau_D \left(\frac{x_c - x_p}{\Delta t}\right) + x_p + \left(\frac{x_c - x_p}{\Delta t}\right) t$$

Now, if t_D and t_N are known constants, and we are given the discrete values of x_p, y_p, and x_c for any given interval Dt, then equation can be solved exactly for the current output value y_c. If the coefficients t_D and t_N are variable, the situation is somewhat more complicated.

We will consider only the case where these coefficients may be treated as functions of time, so that remains linear. In this case the coefficients are specified for a given interval by their initial and final values (i.e., past and current). As with the input function x(t), this in itself is insufficient specification unless the coefficients can be assumed to vary linearly with time between the current instants. Therefore, we make the assumption that

$$\tau_N = \tau_{NP} + \left(\frac{\tau_{NC} - \tau_{NP}}{\Delta t}\right) t$$

$$\tau_D = \tau_{DP} + \left(\frac{\tau_{DC} - \tau_{DP}}{\Delta t}\right) t$$

On this basis, equation can be written

$$\left(\dot{\tau}_D t + \tau_{DP}\right) \frac{dy}{dt} + y = \left(\dot{\tau}_N t + \tau_{NP}\right) \dot{x} + \left(\dot{x} t + x_p\right)$$

where

$$\dot{\tau}_D = \left(\frac{\tau_{DC} - \tau_{DP}}{\Delta t}\right) \quad \dot{\tau}_N = \left(\frac{\tau_{NC} - \tau_{NP}}{\Delta t}\right) \quad \dot{x} = \left(\frac{x_c - x_p}{\Delta t}\right)$$

Knowing the initial and final values of x, t_N, and t_D, and the initial value of y for a given Dt, equation can be solved exactly for the final (current) value of y.

GENERAL FIRST ORDER RECURRENCE FORMULAS

Before deriving the actual solutions of equations we will consider the form these solutions must take. We require a formula that can be applied recursively to compute the current value of y based on the current value of x and the past values of x and y. We will find that, in general, the solution can be expressed as a linear combination of the three given values, i.e.,

$$y_c = (f_1)\, y_p + (f_2)\, x_p + (f_3)\, x_c$$

where f_1, f_2, and f_3 are all functions of t_D, t_N, and Dt. However, these three functions are obviously not independent, because if $y_p = x_p = x_c = m$ then clearly $y_c = m$, so we have

$\mu = \mu f_1 + \mu f_2 + \mu f_3$ which implies $1 = f_1 + f_2 + f_3$. Taking advantage of this fact, we can eliminate f_1 from equation to give

$$y_c = (1 - f_2 - f_3)\, y_p + (f_2)\, x_p + (f_3)\, x_c$$

If we then make the substitution $x_c = x_p + (x_c - x_p)$ we have

$$y_c = (1 - f_2 - f_3)\, y_p + (f_2)\, x_p + (f_3)\, x_p + (f_3)(x_c - x_p)$$

which can be written

$$y_c = y_p + A\,(x_p - y_p) + B\,(x_c - x_p)$$

where the coefficients A and B are defined as

$$A = f_2 + f_3;\ B = f_3$$

Any valid lead/lag algorithm that computes y_c as a linear function of y_p, x_p, and x_c must be expressible in the form of equation, regardless of the assumptions made concerning the interpolation of the independent variable. Therefore, we define equation as the standard recurrence formula for digital lead/lag simulations. In subsequent sections we will frequently identify algorithms simply by specifying their standard recurrence coefficients A and B.

OPTIMUM RECURRENCE COEFFICIENTS FOR CONSTANT τ

When t_D and t_N are constant, the optimum recurrence coefficients can be determined by solving equation, which can be rewritten as follows

$$\frac{dy}{dt}+\left(\frac{1}{\tau_D}\right)y=\left(\frac{\dot{x}}{\tau_D}\right)t+\left(\frac{\tau_N\dot{x}+x_p}{\tau_D}\right)$$

Recall that the solution of any equation of the form

$$\frac{dy}{dt}+F(t)y=G(t)$$

where F(t) and G(t) are functions of t is given by

$$y=\frac{1}{e^r}\left(\int G(t)e^r\ dt+K\right)\text{ where } r=\int F(t)\,dt$$

where K is a constant of integration. Making the substitutions

$$F(t)=\frac{1}{\tau_D}$$

$$F(t)=\left(\frac{\dot{x}}{\tau_D}\right)t+\left(\frac{\tau_D\dot{x}+x_p}{\tau_N}\right)$$

$$r=\int\frac{1}{\tau_D}dt=\frac{t}{\tau_D}$$

we have

$$y=\frac{1}{e^{t/t_D}}\left[\int\left(\frac{\dot{x}}{\tau_D}t+\frac{\tau_D\dot{x}+x_p}{\tau_N}\right)e^{t/t_D}+K\right]$$

Performing the integration and dividing through by $e^{t/tD}$ gives

$$y=\dot{x}t+\dot{x}(\tau_N-\tau_D)+x_p+Ke^{-t/t_D}$$

To determine K we apply the initial condition $y = y_p$ at t = 0, which gives

$$K=y_p-x_p-\dot{x}(\tau_N-\tau_D)$$

Inserting this back into the preceding equation, recalling the definition of x&, and noting that $y = y_c$ at $t = \Delta t$, we arrive (after some algebraic manipulation) at the result

$$y_c=(f_1)y_p+(f_2)x_p+(f_3)x_c$$

where

$$f_1 = e^{-\Delta t/\tau_D}$$

$$f_2 = \frac{\tau_N - \tau_D}{\Delta t}\left(e^{-\Delta t/\tau_D} - 1\right) - e^{-\Delta t/\tau_D}$$

$$f_3 = 1 + \frac{\tau_N - \tau_D}{\Delta t}\left(e^{-\Delta t/\tau_D} - 1\right)$$

Note that as required the sum of f_1, f_2, and f_3 is identically 1. Substituting the expressions for f_2 and f_3 into equations gives the optimum recurrence coefficients for constant t:

$$A = 1 - e^{-\Delta t/\tau_D}$$

$$B = 1 + A\frac{\tau_N - \tau_D}{\Delta t}$$

OPTIMUM RECURRENCE COEFFICIENTS FOR VARIABLE τ

When t_N and t_D are variable, we base the optimum recurrence coefficients on equation. For convenience we define the following parameters

$$a = \dot{\tau}_D$$

$$b = \tau_{DP}$$

$$c = \dot{x}\left(1 + \dot{\tau}_N\right)$$

$$d = \dot{x}\tau_{DP} + x_p$$

In these terms equation can be rewritten

$$\frac{dy}{dt} + \frac{1}{at + b}y = \frac{ct + d}{at + b}$$

This is in the form of equation so the general solution is given by equation where

$$F(t) = \frac{1}{at + b} \quad G(t) = \frac{ct + d}{at + b}$$

$$r = \int \frac{1}{at + b}dt = \text{In}\left((at + b)^{1/a}\right)$$

Therefore we have

$$y = \frac{1}{(at+b)^{1/a}}\left[\int\left(\frac{ct+d}{at+b}\right)(at+b)^{1/a}\,dt + K\right]$$

Performing the integration and dividing through by $(at+b)^{1/a}$ gives

$$y = \left(\frac{c}{1+a}\right)t + d - \frac{bc}{1+a} + K(at+b)^{-1/a}$$

The constant of integration K is determined by the initial condition $y = y_p$ at $t = 0$, which gives

$$K = b^{1/a}\left(y_p - d + \frac{bc}{1+a}\right)$$

We can now compute y_c by evaluating equation at $t = Dt$. If we then replace a,b,c, and d with their respective definitions, we arrive at the result

$$y_c = (f_1)y_p + (f_2)x_p + (f_3)x_c$$

where

$$f_1 = \left(\frac{\tau_{DP}}{\tau_{DC}}\right)^{1/\dot{\tau}_D}$$

$$f_2 = \left[1 - f_1 - \frac{\tau_{NP}}{\Delta t}(1-f_1) + \frac{\tau_{DP}}{\Delta t}\left(\frac{1+\dot{\tau}_N}{1+\dot{\tau}_D}\right)(1-f_1) - \left(\frac{1+\dot{\tau}_N}{1+\dot{\tau}_D}\right)\right]$$

$$f_3 = \left[\frac{\tau_{NP}}{\Delta t}(1-f_1) + \frac{\tau_{NP}}{\Delta t}\left(\frac{1+\dot{\tau}_N}{1+\dot{\tau}_D}\right)(1-f_1) - \left(\frac{1+\dot{\tau}_N}{1+\dot{\tau}_D}\right)\right]$$

Note that, as required, the sum of f_1, f_2, and f_3 is identically 1. Substituting the expressions for f_2 and f_3 into equations gives the optiumum recurrence coefficients for variable τ:

$$A = 1 - \left(\frac{\tau_{DP}}{\tau_{DC}}\right)^{1/\dot{\tau}_D}$$

$$B = \left(\frac{1+\dot{\tau}_N}{1+\dot{\tau}_D}\right) + \frac{A}{\Delta t}\left(\tau_{NP} - \left(\frac{1+\dot{\tau}_N}{1+\dot{\tau}_D}\right)\tau_{DP}\right)$$

When referring to the recurrence formulas the term

"optimum" is justified in the following sense: The total solution y(t) equals the solution to the homogeneous equation plus a particular solution for the given forcing function. The only ambiguity is in the particular solution, which depends on how we choose to interpolate the independent function x(t). Note that the forcing function has no effect on the homogeneous solution, and that the particular solution is independent of the actual y(t) values.

It follows that the coefficients of the "y terms" in the recurrence relation must be exactly, *regardless of how the x(t) function is interpolated*. The only ambiguityt in the recurrence formula is in the "x-term" coefficients (f_2 and f_3), and even those have a fully determined sum. In view of this, the term "optimum" is used throughout this document to denote a recurrence formula that has the exact "y-term" coefficient(s), and for which the "x-term" coefficients sum to the correct value.

With regard to the variable-t solution, we would expect to find that it has as a special case the constant-t solution, and in fact if and are zero then clearly equation is equivalent to the constant-t expression for B given by equation. However, it may not be self-evident that equation reduces to the A of equation for constant t. To show that it does, we can rewrite equation as follows $A = 1 - e^{(1/\tau_D)\ln(\tau_{DP}/\tau_{DC})}$

If we now recall the series expansion of the natural log

$$\ln(z) = 2\left[\left(\frac{z-1}{z+1}\right) + \frac{1}{3}\left(\frac{z-1}{z+1}\right)^3 + \ldots\right]$$

we see that if the ratio t_{DP}/t_{DC} is close to 1 we can approximate the natural log in equation very accurately by just the first term of the expansion, i.e.,

$$\ln\left(\frac{\tau_{DP}}{\tau_{DC}}\right) \approx 2\left[\frac{\left(\frac{\tau_{DP}}{\tau_{DC}}\right) - 1}{\left(\frac{\tau_{DP}}{\tau_{DC}}\right) + 1}\right] = 2\left(\frac{\tau_{DC} - \tau_{DP}}{\tau_{DC} + \tau_{DP}}\right)$$

in which case equation becomes

$$A \approx 1 - e^{\frac{\Delta t}{(\tau_{DP}+\tau_{DC})^2}}$$

This makes it clear that as $\dot{\tau}_D$ goes to zero, and t_{DC} approaches t_{DP}, the vairable-t expression for A does in fact reduce to the constant-t case given by equation.

We now demonstrate that the constant-t response approaches the variable-t response if a sufficiently small time interval Δt is used. First, notice that as Δt goes to zero the ratio of t_{DP} to t_{DC} can be made arbitrarily close to 1 for any finite value of $\dot{\tau}_D$. Also, as we have already seen, equation is equivalent to A of equation in the limit as τ_{DP}/τ_{DC} goes to 1. Therefore, as Δt goes to zero the A coefficient for both constant t and variable t is given by equation. Furthermore, if we recall the power series expansion

$$e^z = 1 + \frac{z}{1!} + \frac{z^2}{2!} + \frac{z^3}{3!} + \ldots$$

we see that as Δt goes to zero, A of equation becomes simply $A = \Delta\tau/\tau_D$. Substituting this into the expressions for B in equations, along with the stipulation that $\tau_{NC} \approx \tau_{NP}$ for a sufficiently small Dt, we see that both solutions give $B = t_N/t_D$. Thus, for sufficiently small Δt, the constant-τ and variable-τ solutions both reduce to the form

$$y_c = y_p + \frac{\Delta t}{\tau_D}\left(x_p - y_p\right) + \frac{\tau_N}{\tau_D}\left(x_c - x_p\right)$$

Notice that if Δt is actually zero, but $x_c - x_p$ does not vanish, then this can be written as

$$y_c = y_p + \frac{\tau_N}{\tau_D}\left(x_c - x_p\right)$$

which is the expected response to a step input, viz, an instantaneous change in x yields an instantaneous change in y with a magnitude amplified by the factor τ_N/τ_D.

Examination of equation also shows that the variable-t solution requires t_{DP} and t_{DC} have the same sign, since if they didn't, the ratio τ_{DP}/τ_{DC} would be negative and the result of the exponentiation would, in general, be complex. Another way of stating this restriction is that τ_D can never pass through zero, which effectively prohibits a change of sign for a continuous function. In one sense, the "reason" for this restriction is that we divided by t_D when we wrote equation. More fundamentally, equation shows that when τ_D is zero the differential term in y vanishes and the equation is singular.

It may appear that equation also exhibits a singularity, specifically when $\dot{\tau}_D$ Δt equals -1. However, as long as the absolute value of $\Delta t\left(\dot{\tau}_D / \tau_{DP}\right)$ is less than 1 (which corresponds to the requirement that τ_D never pass through zero during the interval) it can be shown that B remains analytic at $\dot{\tau}_D = -1$, and is given by

$$B = \frac{\tau_{NP}}{\tau_{DP}} + \left(1 + \dot{\tau}_N\right)\left[1 - \left(1 - \frac{\tau_{DP}}{\Delta t}\right) \ln\left(1 - \frac{\Delta t}{\tau_{DP}}\right)\right]$$

FINITE DIFFERENCES AND BILINEAR SUBSTITUTION

The most direct approach to determining an approximate recurrence formula for a differential equation of the form $\tau_D \frac{dy}{dt} + y = \tau_N \frac{dx}{dt} + x$ is to substitute simple finite differences for the differentials about the nearest convenient point. The derivative of the function y(t) at the point half-way between the past and current instants can be approximated by

$$\frac{dy}{dt} \approx \frac{y_c - y_p}{\Delta t}$$

The value of the function y(t) at this point is approximately

$$y \approx \frac{y_c + y_p}{2}$$

Similarly for the function x(t) at the same instant we have the approximations

$$\frac{dx}{dt} \approx \frac{x_c - x_p}{\Delta t} \qquad x \approx \frac{x_c + x_p}{2}$$

Making these substutitions into equation and rearranging gives the recurrence formula

$$y_c \approx \left(\frac{2\tau_D - \Delta t}{2\tau_D + \Delta t}\right) y_p + \left(\frac{-2\tau_N + \Delta t}{2\tau_D + \Delta t}\right) x_p + \left(\frac{2\tau_N + \Delta t}{2\tau_D + \Delta t}\right) x_c$$

By equations, the standard form recurrence coefficients are

$$A = \frac{\Delta t}{\tau_D + \Delta t / 2} \quad B = \frac{\tau_N + \Delta t / 2}{\tau_D + \Delta t / 2}$$

These are identical to equations, which shows that the direct application of finite differences gives the same first-order recurrence formula as results from substituting the exponential approximation of equation into equations.

There is another approach that, called bilinear substitution (or Tustin's algorithm), that leads to the same result as finite differences (and the optimum exponential substitution). To apply this method we first re-write equation in the form

$$(\tau_D D + 1) y_c = (\tau_N D + 1) y_c$$

where D is the differential operator. We then make the substitution

$$D = \frac{2}{\Delta t}\left(\frac{1 - z^{-1}}{1 + z^{-1}}\right)$$

where the symbol z^{-1} is the "past value" operator. For example, $z^{-1}\, y_c = y_p$. Substituting equation into equation, we have

$$\left[\frac{2\tau_D}{\Delta t}\left(\frac{1 - z^{-1}}{1 + z^{-1}}\right) + 1\right] y_c = \left[\frac{2\tau_N}{\Delta t}\left(\frac{1 - z^{-1}}{1 + z^{-1}}\right) + 1\right] x_c$$

Multiplying through by $\Delta t(1+z^{-1})$ gives

$$\left[2\tau_D\left(1 - z^{-1}\right) + \Delta t\left(1 + z^{-1}\right)\right] y_c = \left[2\tau_N\left(1 - z^{-1}\right) + \Delta t\left(1 + z^{-1}\right)\right] x_c$$

Rearranging terms, we have

$$(2\tau_D + \Delta t)\, y_c + (2\tau_D + \Delta t)\, z^{-1} y_c = (2\tau_N + \Delta t)\, x_c + (2\tau_N + \Delta t)\, z^{-1} x_c$$

Carrying out the indicated past value operations gives

$$(2\tau_D + \Delta t)\, y_c \pm (2\tau_D + \Delta t)\, y_p = (2\tau_N + \Delta t)\, x_c + (2\tau_N + \Delta t)\, x_p$$

Solving for y_c, we arrive at equation. Thus, when applied to a first-order transfer function, the bilinear substitution method gives the same recurrence formula as the method of finite differences.

PRE-WARPING

A technique called "pre-warping" is sometimes applied to lead/lag algorithms in an attempt to correct the amplitude of the output responding to an oscillating input signal. The standard recurrence coefficients for this method are

$$A = \frac{2}{1 + \dfrac{1}{\tan(\Delta t / 2\tau_D)}} \qquad B = \frac{1 + \dfrac{1}{\tan(\Delta t / 2\tau_N)}}{1 + \dfrac{1}{\tan(\Delta t / 2\tau_D)}}$$

Recalling that as z goes to zero, tan(z) approaches z, we see that the pre-warped coefficients reduct to equations if Δt is small relative to τ_N and τ_D.

The rationale for pre-warping is clarified by the following example. Given a lead/lag simulation with time-step size Δt = 0.25 seconds and a sinusoidal input function with a period of T = 1 second, each cycle of the input will be described by four discrete values.

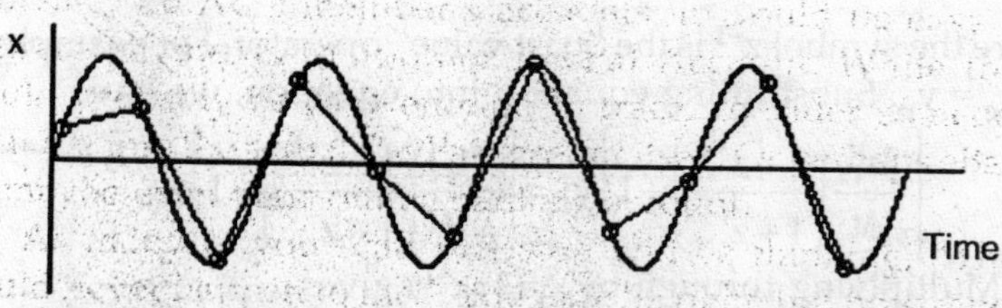

Clearly a sample will rarely be taken at an absolute peak of the wave, even if T is not an exact multiple of Δt and the samples precess through the cycle. Therefore, assuming linear interpolation between the discrete values, we will truncate most of the peaks and under-represent the true amplitude of the input function, which results in a corresponding underestimation of the output amplitude. In an attempt to correct for this, pre-warping biases the recurrence coefficients to reduct the effective lag of the filter, thereby increasing its output amplitude for a given oscillating input. However, the bias necessary to cancel out the under-representation of the input varies with the ration of T to Dt, so the pre-warped coefficients are really only correct for one particular frequency - by convention, the "half power frequency". At lower frequencies, and for most non-oscillatory inputs, the pre-warping coefficients yield less accurate results than the approximate coefficients given by equations. Hence, pre-warping is a highly specialized technique, intended only for a limited range of frequency-response applications.

COMPOSITE PAST VALUE RECURRENCE

All the solution methods discussed so far have assumed the availability of xp and yp, that is, past values of the input and out functions. This requires the dedication of two RAM memory locations for each lead/lag in the system. In contrast, the composite past value (CPV) recurrence scheme required only one stored value. In the constant-t case this recurrence scheme is functionally equivalent to the standard recurrence formula for any given solution coefficients A and B. However, when CPV recurrence is applied in the variable-t case, it does not give exactly the same results as the standard formula.

To derive CPV recurrence formula, we begin by rewriting the standard formula as follows

$$y_c = (1 - A)\, y_p + (A - B)\, x_p + (B)\, x_c$$

Dividing and multiplying the two past-value terms by 1–B gives

$$y_c = (1 - B)\left[\frac{1-A}{1-B}\, y_P + \frac{A-B}{1-B}\, x_P\right] + (B)x_c$$

If we define the quantity in the brackets as the "composite past value", denoted by z_p, then we can rewrite equation as

$$y_c = (1 - B)\, z_p + (B)\, x_c$$

or

$$y_c = z_p + B\,(x_c - z_p)$$

The composite *current* value is defined as

$$z_c = \frac{1-A}{1-B} y_c + \frac{A-B}{1-B} x_c$$

To simplify this, we substitute the expression for the y_c from equation, which gives

$$z_c = (1-A) z_p + \frac{B(1-B)}{1-B} x_c + \frac{A-B}{1-B} x_c$$

Combining terms in x_c and simplifying, we have

$$z_c = z_p + A(x_c - z_p)$$

Therefore, the CPV recurrence formulas are

$$y_c = z_p + B\,(x_c - z_p)$$
$$z_c = z_p + A\,(x_c - z_p)$$

The derivation of these formulas does not distinguish between the past and current values of A and B, thereby tacitly assuming that these coefficients are constant. However, when τ_N and τ_D are variable, the standard coefficients A and B are also variable, in which case the CPV recurrence formulas are not exactly equivalent to the standard formula on which they are based. To demonstrate this, we can solve equation for z_p and substitute this into equation to give

$$z_c = \frac{1-A_c}{1-B_c} y_c + \frac{A_c - B_c}{1-B_c} x_c$$

A and B are now subscripted with a "c" to denote the current values, since only the current values are available each time the recurrence formulas are executed. Equation is also valid at the past instant, which is to say

$$z_p = \frac{1-A_p}{1-B_p} y_p + \frac{A_p - B_p}{1-B_p} x_p$$

If we substitute this expression for z_p into equation, the resulting equation for y_c can be reduced to the form

$$y_c = y_p + A_{eff}\,(x_p - y_p) + B_{eff}\,(x_c - x_p)$$

where A_{eff} and B_{eff} are the effective standard recurrence coefficients, related to the actual coefficients used in the CPV recurrence formula by

$$A_{eff} = \left(\frac{1-B_c}{1-B_P}\right)A_p + \left(\frac{B_c - B_P}{1-B_P}\right) \quad B_{eff} = B_c$$

Because of this effect, there are circumstances in which CPV recurrence yields unrealistic output response, even if the optimum formulas for A and B are used.

The simplest illustration of this is the case where τ_N and τ_D initially have different values and then become equal. The true first order response is for the output to asymptotically converge on the input function once $\tau_N = \tau_D$, and this can be accuratly simulated using the standard recurrence formula. However, this response cannot be reproduced using CPV recurrence.

To understand why, notice that whenever $\tau_N = \tau_D$ the current value of B (denoted by B_c) must equal 1, as is clearly shown by equation. Substituting this into equations gives the effective coefficients $A_{eff} = 1$ and $B_{eff} = 1$. Inserting these into the standard recurrence formula gives the result

$$y_c = y_p + (1)(x_p - y_p) + (1)(x_c - x_p) = x_c$$

Thus, using CPV recurrence, the output undergoes a "step" change and "instantaneously" becomes equal to the current input, regardless of the output's past value.

A possible objection to equation -1) is that it appears to predict divergent behaviour when B_p equals 1, whereas in practive CPV recurrence formulas never yield divergent response. The explanation is that B_p can only equal 1 if B_c of the previous iteration was 1.

But we have already seen that if B_c was 1 for the previous iteration, then y_c was unconditionally set equal to x_c during that iteration. Therefore, for the present iteration, y_p must equal x_p. Recalling that A is the coefficient $(x_p - y_p)$ in the standard recurrence formula, we see that, in effect, the divergence of A_{eff} at $B_p = 1$ is always "cancelled" by the vanishing of $(x_p - y_p)$.

Digital simulations of continuous dynamic functions have a constraint on their execution frequency imposed by the intrinsic information rate of the input and output signals. The execution frequency must be high enough to resolve and convery the significant characteristics of these signals. However, the simplified methods of simulation described in Sections have another limitation:

The approximations on which they are based are valid only if the execution period Δt is small relative to the characteristic times of the function (i.e., t_N and t_D). To illustrate that these two limitations are logically independent, consider the simple lag response

$$\tau\frac{dy}{dt} + y = x$$

where t = 0.05 seconds. Equations give the recurrence formula

$$y_c = y_p + \left(\frac{2\Delta t}{0.1 + \Delta t}\right)(x_p - y_p) + \left(\frac{\Delta t}{0.1 + \Delta t}\right)(x_c - x_p)$$

Now suppose that, based on their maximum information rates, we have determined that x and y can be adequately represented with a sampling frequency of 4 Hz, corresponding to a Δt of 0.25 seconds. Substituting this into equation gives

$$y_c = y_p + (1.4286)(x_p - y_p) + (0.7143)(x_c - x_p)$$

A sample of the response given by this recurrence formula for a single time-step is illustrated.

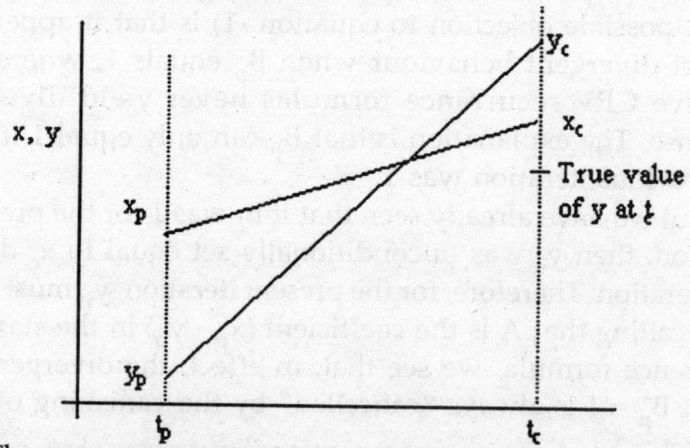

The output computed by equation crosses over the input, which is obviously not representative of a pure lag response. In general, recurrence formulas based on equations will not give an accurate representation of a lead/lag function unless

Δt is less than 2t. This means that the execution frequency of the present example would have to be raised from 4 Hz to at least 10 Hz. In contrast, if the optimum coefficients given by equations are used, the recurrence formula is

$$y_c = y_p + (0.9933)(x_p - y_p) + (0.8013)(x_c - x_p)$$

for which an execution frequency of 4 Hz is adequate. Thus the use of non-optimal coefficients imposes unnecessarily high execution frequency on the simulation.

Another way of looking at this is to regard the error in the computed value of y_c for any given interval Δt as consisting of two parts: one part due to the uncertainty in the input function, and one part resulting from the approximations made in determining the coefficients of the recurrence formula. The first part is unavoidable (for a given Δt), but he second part can be eliminated entirely by using the optimum recurrence coefficients.

With the technique of pre-warping, an attempt is made to use these two errors to compensate for each other. These two errors offset each other only under a limited range of conditions. Essentially, pre-warping attempts to account for non-inearity in the input signal over the interval Dt, but there is really no general way of determining the non-linearity of a signal based on just two samples (x_p and x_c). In applying its "correction", pre-warping makes use of information about the input signal beyond what can be inferred from the two given points.

Specifically it depends on the knowledge that the input is periodic with a given frequency. Without such special information, a general method of accounting for non-linearity of the input signal must make use of at least three values of x. With three values we can determine the optimum recurrence formula based on second order interpolation of the input function. Denoting the "past past value" of x by x_{pp}, the expanded recurrence formula can be written

$$y_c = (g_1)\, y_p + (g_2)\, x_{pp} + (g_3)\, x_p + (g_4)\, x_c$$

where

$$g_1 = e^{1\Delta t/\tau_D}$$

$$g_2 = \frac{\tau_N - \tau_D}{2\Delta t}\left[\left(1 - \frac{2\tau_D}{\Delta t}\right) + e^{-\Delta t/\tau_D}\left(1 + \frac{2\tau_D}{\Delta t}\right)\right]$$

$$g_3 = \frac{2(\tau_N - \tau_D)}{\Delta t}\left[\frac{\tau_D}{\Delta t}\left(1 - e^{-\Delta t/\tau_D}\right) - 1\right] - e^{-\Delta t/\tau_D}$$

$$g_4 = 1 + \frac{\tau_N - \tau_D}{2\Delta t}\left[\left(1 - e^{1\Delta/\tau_D}\right) + 2\right]$$

By the same reasoning that led to the relation $f_1 + f_2 + f_3 = 1$, we have the relation

$$g_1 + g_2 + g_3 + g_4 = 1$$

Using this, we can re-write equation in a generalized standard form analagous to equation as follows

$$y_c = y_p + A\,(x_p - y_p) + B\,(x_c - x_p) + C\,(x_c - 2x_p + x_{pp})$$

where

$$A = 1 - e^{-\Delta t/\tau_D}$$

$$B = 1 + A\left(\frac{\tau_N - \tau_D}{\Delta t}\right)$$

$$C = \frac{(\tau_N - \tau_D)}{\Delta t}\left[A\left(\frac{1}{2} + \frac{\tau_D}{\Delta t}\right) - 1\right]$$

If x_{pp}, x_p, and x_c are co-linear vs time, then the quantity multiplied by C vanishes, and equation reduces to equation.

Beginning with equation), we could have defined the composite past value directly as just the sum of the past value terms

$$z_p = (1 - A)\,y_p + (A - B)\,x_p$$

Then the CPV recurrence formulas would have been

$$y_c = z_p + B\,x_c$$
$$z_c = (1 - A)\,y_c + (A - B)\,x_c$$

The advantages of equations is that they can be executed with just two multiplications, whereas equations and involve three multiplications. This is significant because multiplication is a time-consuming operation in most digital devices. However, equations will give better accuracy in the variable-t case because they are equivalent to using the effective standard coefficients

$$A_{eff} = A_p + (B_c - B_p)$$
$$B_{eff} = B_c$$

These coefficients never yield singular values of the coefficients, unlike equations when B_c=1, so no special provisions are needed to account for this computational singularity (recognizing that it cancels when multiplied by $(x_c - x_p)$).

Chapter 9

Circular Distributions

INTRODUCTION

Consider a circular arrangement of D bins, into which are placed n marbles randomly with uniform distribution. What is the probability that all the marbles will be located within N consecutive bins? If N is less than or equal to D/2, the answer is easy to see, as discussed in another note.

Briefly, the question is equivalent to whether n randomly placed arc spans of length N all have some common overlap, and we can consider how many of the D^n different arrangements of n such arcs intersect in at least one bin. Focusing on just one particular bin, we know there are N^n distinct arrangements for which every arc includes that bin.

Multiplying this by D, the number of bins, gives a first estimate of the total number of arrangements with at least one bin in common. Each arrangement with exactly one bin in common has been counted exactly once in this estimate, but each arrangement with two bins in common has been counted twice (once for each of the common bins).

Likewise each arrangement with three common bins has been counted three times, and so on. To eliminate the duplicates, consider the number of arrangements that have (at least) two particular adjacent bins in common.

The number of such arrangements for any particular pair of adjacent bins is clearly $(N-1)^n$, and we can again multiply this by D to give an estimate of the total number of such arrangements. This counts the arrangements with exactly two common bins just once, but it counts those with three common bins twice, and so on. Therefore, subtracting this number from the previous number, and dividing the result by D^n (the total number of all possible arrangements) gives the probability of overlap between all n of the arcs

$$\Pr\{n, N, D\} = \frac{D(N^n - (N-1)^n)}{D^n}$$

But, again, this applies only if N is less than or equal to D/2. The question is how to generalize this for cases when N is greater than D/2. The solution for continuous circular arcs is well known, but the discrete case seems to be less so. At one extreme, with N = D–1, the answer is easy to see. Letting c_j denote the event that none of the marbles is located in the jth cell (bin), the event of interest is the union of these n events, so the probability of all n marbles being located in an arc of length N = D-1 is $\text{pr}\{c_1 + c_2 + \ldots + c_D\} =$

$$\sum_{j=1}^{D} \Pr\{c_j\} - \sum_{j=1}^{D-1} \Pr\{c_j c_k\} + \sum_{j=1}^{D-2} \sum_{k=j+1}^{D-1} \sum_{j=1}^{D} \Pr\{c_j c_k c_r\} - \ldots.$$

Hence the probability is

$$\Pr\{n, D-1, D\} = \sum_{j=1}^{D} (-1)^{j-1} \binom{D}{j} \left(\frac{D-j}{D}\right)^n$$

In the case N = D – 2 we need to assess the probability of the event

$$c_1 c_2 + c_2 c_3 + c_3 c_4 + \ldots + c_{D-1} c_D + c_D c_1$$

The first-order terms are all products of two distinct elements, and there are D such terms, so they contribute $D[(D-2)/D]^n$ to the probability, but the second-order terms are more complicated, because the first-order terms are not all mutually independent. There are a total of D(D–1)/2 second-order terms, and of these, D consist of products of three distinct elements, and D(D-3)/2 consist of products of four distinct elements.

Proceeding to the third order terms, there are a total of D(D-1)(D-2)/6, consisting of D products of four, D(D–4) products of five, and D(D-4)(D-5)/6 products of six distinct elements. We can continue to find higher order terms, until all combinations have been exhausted, and then consolidate terms, noting that the kth order terms consisting of j products contribute

$$(-1)^{k-1}\left(\frac{D-j}{D}\right)^n$$

to the total probability. The same approach can be taken for any N, by setting the D basic elements to the D products of D–N consecutive elements c_i. As an example, suppose we wish to determine the formula for the probability of all n marbles lying within a span of N=12 cells in a circle of D=16 cells. We seek the probability of the event

$$c_1c_2c_3c_4 + c_2c_3c_4c_5 + c_3c_4c_5c_6 + \ldots.$$
$$\ldots + c_{13}c_{14}c_{15}c_{16} + c_{14}c_{15}c_{16}c_1 + c_{15}c_{16}c_1c_2 + c_{16}c_1c_2c_3$$

Naturally, the sum of the terms on the kth row is the binomial coefficient C(16,k). From this table we can infer that, given a circle of 16 cells, the probability of all n marbles lying within 12 consecutive cells is

$$\Pr\{n,12,16\} = 16\left[\left(\frac{12}{16}\right)^n - \left(\frac{12}{16}\right)^n\right] - 56\left[\left(\frac{8}{16}\right)^n - 2\left(\frac{7}{16}\right)^n + \left(\frac{6}{16}\right)^n\right]$$
$$+ 16\left[\left(\frac{4}{16}\right)^n - 3\left(\frac{3}{16}\right)^n + 3\left(\frac{2}{16}\right)^n - \left(\frac{1}{16}\right)^n\right]$$

In the same way, we can determine the probabilities with smaller values of D but the same ratio of N/D. This gives

$$\Pr\{n,3,4\} = 4\left[\left(\frac{3}{4}\right)^n - \left(\frac{2}{4}\right)^n\right] - 2\left[\left(\frac{2}{4}\right)^n - 2\left(\frac{1}{4}\right)^n + \left(\frac{0}{4}\right)^n\right]$$

$$\Pr\{n,6,8\} = 8\left[\left(\frac{6}{8}\right)^n - \left(\frac{5}{8}\right)^n\right] - 12\left[\left(\frac{4}{8}\right)^n - 2\left(\frac{3}{8}\right)^n + \left(\frac{2}{8}\right)^n\right]$$

$$\Pr\{n,9,12\} = 12\left[\left(\frac{9}{12}\right)^n - \left(\frac{8}{12}\right)^n\right] - 30\left[\left(\frac{6}{12}\right)^n - 2\left(\frac{5}{12}\right)^n + \left(\frac{4}{12}\right)^n\right]$$

To see the correspondence with the continuous solution, consider the first terms of these expressions, which are multiples of quantities of the form

$$(2q-1)^n - 2(2q-1-\alpha)^n + (2q-1-2\alpha)^n = \frac{2\alpha^2}{2q-1}\binom{n}{2}(2q-1)^{n-1} + \dots$$

where we have put $q = N/D$ and a = 1/D, and we have expanded the expression in powers of a. The coefficient of each of these quantities is simply $D = 1/\alpha$, so in the limit as D goes to infinity (and a goes to zero) these terms represent simply the quantity nq^{n-1}.

At this state it would be tempting to think that the coefficients of these terms are

$$?\ \frac{2q-1}{2\alpha^2} = \frac{D}{2}(2N-D)\ ?$$

but in fact to give agreement with the preceding results we must subtract D/2, and we can clearly add or subtract any constant multiple of D/2 without affecting the limiting value, since the ratio of (2N-*D*–*K*)/(2*N*–*D*) for any constant *K* still goes to 1 as *N* and *D* increase to infinity. Thus the actual coefficients of the second terms are given by

$$\frac{D}{2}(2n-d-1) = \frac{D}{2N-D}\binom{2N-D}{2}$$

The third terms (if present) are multiples of quantities of the form

$$(3q-2)^n - 3(3q-2-\alpha)^n + 3(3q-2-2\alpha)^n - (3q-2-3\alpha)^n$$

$$= \frac{6\alpha^3}{(3q-2)^2}\binom{n}{3}(3q-2)^{n-1} + \dots$$

Again the temptation is to assume the coefficients of these terms are the reciprocal of the leading factor in this expression, i.e.,

$$?\ \frac{(3q-2)^2}{6\alpha^3} = \frac{D}{6}(3N-2D)^2\ ?$$

But again, in order to agree with the coefficients derived by inclusion-exclusion, we must deduct a constant from each of the factors in the square. In this case we subtract 1 from the first and 2 from the second, giving the coefficients of the third terms as

$$\frac{D}{6}(3N-2D-1)(3N-2D-2)=\frac{D}{3N-2D}\binom{3N-2D}{3}$$

The pattern for higher order coefficients is clear, and of course for any given N and D we will have only k non-zero terms, where kN-(k–1)D must be greater than or equal to k. Hence the number of terms is the greatest integer less than or equal to D/(D–N+1). This explains why, if N is less than or equal to D/2, there is only one term (as we discussed originally). With these coefficients, the continuous case for n points on a circle all lying within an arc of q times the whole perimeter is given by the well-known formula

$$\Pr\{n,q\}=\sum_{j=1}^{j_{\max}}(-1)^{j-1}\binom{n}{j}[1-j(1-q)]^{n-1}$$

where j_{max} is the greatest integer less than but *not* equal to 1/(1–q). In our previous example we had N/D = q = 3/4, so j_{max} = 3, and this formula gives

$$\Pr\left\{n,\frac{3}{4}\right\}=\sum_{j=1}^{3}(-1)^{j-1}\binom{n}{j}\left[1-\frac{j}{4}\right]^{n-1}=\binom{n}{1}\left(\frac{3}{4}\right)^{n-1}$$
$$-\binom{n}{2}\left(\frac{2}{4}\right)^{n-1}+\binom{n}{2}\left(\frac{1}{4}\right)^{n-1}$$

which agrees exactly with the result of taking the limit of the discrete cases with N/D = 3/4 as D goes to infinity.

$$\Pr\{n,N,D\}=\sum_{j=1}^{j_{\max}}\left[\frac{(-1)}{jN-(j-1)D}\binom{jN-(j-1)D}{j}\sum_{k=0}^{j}(-1)^k\binom{j}{k}\left(\frac{jN-(j-1)D-k}{D}\right)^n\right]$$

where j_{max} is the greatest integer less than or equal to D/(D-N+1). Letting M denote D-N, this can be written more succinctly as

$$\Pr\{n, N, D\} = \sum_{j=1}^{j_{\max}} \left[\frac{(-1)^{j-1} D}{D - Mj} \binom{D - Mj}{j} \sum_{k=0}^{j} (-1)^k \binom{j}{k} \left(\frac{D - Mj - k}{D} \right)^n \right]$$

To illustrate the application of this formula, suppose there is a circular arrangement of D = 26 cells, and n marbles are distributed randomly (with uniform density) into these cells.

To determine the probability that all n marbles will lie within a span of $N = 19$ cells, we have $j_{\max} = 3$, because this is the greatest integer less than or equal to 26/(26-19+1).

We also have $M = 26\text{-}19 = 7$, so the probability can be expressed as

$$\Pr\{n, 19, 26\} = \sum_{j=1}^{3} \left[(-1)^{j-1} \frac{26}{26 - 7j} \binom{26 - 7j}{j} \sum_{k=0}^{j} (-1)^k \binom{j}{k} \left(\frac{26 - 7j - k}{26} \right)^n \right]$$

$$\Pr\{n, 19, 26\} = \sum_{j=1}^{3} \left[(-1)^{j-1} \frac{26}{26 - 7j} \binom{26 - 7j}{j} \sum_{k=0}^{j} (-1)^k \binom{j}{k} \left(\frac{26 - 7j - k}{26} \right)^n \right]$$

$$= 26 \left[\left(\frac{19}{26} \right)^n - \left(\frac{18}{26} \right)^n \right]$$

$$-143 \left[\left(\frac{12}{26} \right)^n - 2 \left(\frac{11}{26} \right)^n + \left(\frac{10}{26} \right)^n \right]$$

$$+52 \left[\left(\frac{5}{26} \right)^n - 3 \left(\frac{4}{26} \right)^n + 3 \left(\frac{3}{26} \right)^n - \left(\frac{2}{26} \right)^n \right]$$

DESCRIPTION

Then four OTHER numbers are arbitrarily drawn from the list (without replacement) and given to us, and our task is to communicate knowledge of the winning number to someone else simply by sending them our four numbers arranged in some particular order. It's easy to establish a protocol for communicating this information on this basis, because the recipient knows the winning number is not one of the four he receives from us, so there are really only 24 possibilities, which can be arranged in ascending order and placed in one-to-one correspondence with the 24 distinct permutations of the four numbers designated a,b,c,d in ascending order.

But what if we relax our protocol condition slightly, and permit ourselves to send either all four of the message-numbers in some order OR to send just three of the numbers in some order?

In this case we can't rule out the possibility of an effective protocol, because the number of distinct answers we can give with a fixed set of four message-numbers is 24 permutations of four PLUS 6 permutations of three, and there are four distinct ways of choosing three of the four message numbers.

One protocol that accomplishes our objective is based on a predefined set of 6-element subsets of the original 42 numbers, defined so as to minimize the overlaps between subsets.

For example, we can take the 42 columns from the arrays shown below:

42 41 40 39 38 37 36	35 34 33 32 31 30 29	28 27 26 25 24 23 22
1 2 3 4 5 6 7	1 2 3 4 5 6 7	1 2 3 4 5 6 7
8 9 10 11 12 13 14	9 10 11 12 13 14 8	10 11 12 13 14 8 9
15 16 17 18 19 20 21	17 18 19 20 21 15 16	19 20 21 15 16 17 18
22 23 24 25 26 27 28	25 26 27 28 22 23 24	28 22 23 24 25 26 27
29 30 31 32 33 34 35	33 34 35 29 30 31 32	30 31 32 33 34 35 29
36 37 38 39 40 41 42	41 42 36 37 38 39 40	39 40 41 42 36 37 38
21 20 19 18 17 16 15	14 13 12 11 10 9 8	7 6 5 4 3 2 1
1 2 3 4 5 6 7	1 2 3 4 5 6 7	1 2 3 4 5 6 7

11 12 13 14 8 9 10	12 13 14 8 9 10 11	13 14 8 9 10 11 12
21 15 16 17 18 19 20	16 17 18 19 20 21 15	18 19 20 21 15 16 17
24 25 26 27 28 22 23	27 28 22 23 24 25 26	23 24 25 26 27 28 22
34 35 29 30 31 32 33	31 32 33 34 35 29 30	35 29 30 31 32 33 34
37 38 39 40 41 42 36	42 36 37 38 39 40 41	40 41 42 36 37 38 39

The first array simply lists the numbers 1 through 42 consecutively in horizontal rows. The second array shifts the nth row $1(n-1)$ places to the right (with wrap-around). The third array shifts the nth row 2(n-1) places to the right, and so on. Each column is associated with a unique number from 1 to 42, noted above it.

Our transmission protocol is based on four given "message-numbers" which we will call a,b,c,d in ascending order. There is also a single winning number w. We check each of the four possible sets of three message numbers, (a,b,c), (a,b,d), (a,c,d), and (b,c,d) and compute the sum of the three numbers, and subtract 42 if necessary to place the reduced sum in the range from 1 to 42. (We could more easily express this if we made the range 0 to 41 and worked modulo 42). This will necessarily give us four distinct indices, because no two of these triples differ by more than one number, and the basic elements are distinct.

We then examine the columns in the above arrays corresponding to those four indices. If the winning number w appears in one of those columns, we then transmit the corresponding triple, with its elements arranged in one of the six permutations to pick out the appropriate number from the indexed column.

If, on the other hand, the winning number is NOT in any of the four indexed columns, we will transmit all four message numbers.

Of course, these can only be arranged in 24 distinct ways, but these suffice, because we can rule out all the numbers in the four indexed columns, as well as the four message-numbers themselves.

If there were no overlap, this would exclude 28 numbers, leaving only 14 to be selected by the 24 permutations.

However, there is typically overlap between the numbers in the four indexed columns, and between the column numbers and the message numbers. We can assign the column indices so that no index appears in its own column (to do this in the above arrays we would make the column transpositions (12,13), (20,21), (27,28), (34,35), and (38,39)), but there can still be overlap from one column to another index.

Nevertheless, we are assured that the fours columns (even without the four given incides) will always cover AT LEAST 18 distinct numbers, because every pair of columns shares no more than one number in common, so the most overlap would occur if each of the six pairs of four columns shared a number, thereby reducing the total from 24 down to 24-6 = 18.

Hence, from the overall 42 numbers, we are left with no more than 42-18 = 24 possible values of w after excluding those that appear in the four columns tat can be indexed by three of the four given message-numbers.

These can therefore be arranged in order and placed in unique correspondence with the permutations of the four message numbers.

ARRANGEMENTS OF REGULAR POLYGONAL

We can distribute six consecutive integers, such as 0 to 5, on the vertices and edges of a triangle in such a way that the sum along each edge is the same. In fact, there are four essentially distinct such arrangements, shown below.

2	4	5	5
3 4	1 3	0 2	0 1
1 5 0	2 5 0	3 4 1	4 2 3

The common sums of these four triangular arrangements are 6, 7, 8, and 9 respectively. Of course, each of these arrangements has six versions if we count rotations and reflections.

Similarly, arrangements of any 8 consecutive numbers can be constructed on the vertices and edges of a square. In this case there are six essentially distinct arrangements, illustrated below.

5 3 1	4 1 5	3 6 2	7 0 5
4 6	6 2	1 5	3 1
0 7 2	0 7 3	7 0 4	2 4 6
	7 2 1	7 1 3	
		3 5	4 2
		0 6 4	0 5 6

The common sums of the squares in each of these four columns are 9, 10, 11, and 12 respectively. It's easy to show algebraically that, for each square, the sums of opposite pairs of edges must be equal. If we let V and E denote the sums of the values on the vertices and edges respectively, then the common sum is S = (E+2V)/n for an n-sided polygon. We also know that there are a total of $2n$ vertices and edges, so we are distributing the integers 0 to 2n-1, which implies that the sum of all the numbers is $E + V = n(2n-1)$. Solving these equations for E and V, we have

$$E = n[2(2n-1)-S] \quad V = n[S-(2n-1)]$$

This shows that both E and V must be divisible by n, and also that the common sum S must be greater than $(2n-1)$ and less than 2(2n-1).

In fact, since there are n edges, each containing n distinct non- negative integers, the least possible value of E is $n(n-1)/2$, so the quantity $2(2n-1)-S$ must not only be positive, it must be greater than or equal to (n–1)/2, and thus S must be less than or equal to $2(2n-1)-(n-1)/2$ = (7n-3)/2. Likewise V must be at least $n(n-1)/2$, so S cannot be less than $(2n-1) + (n-1)/2$ = (5n–3)/2. For $n = 4$ this means S cannot be less than 8.5 and cannot exceed 12.5, consistent with the fact that the smallest common sum is 9 and the largest is 12.

We can place 10 consecutive numbers on the vertices and edges of a pentagon such that the three numbers along any edge have the same sum.

This can be done in six essentially distinct ways. Rather than drawing the figures, we can just list the strings of numbers around the perimeter of the figure, beginning with a vertex. The six arrangements are listed below, along with their common sum.

Common sum		
11	1738092546	
13	0765283194	6523814907
14	3476185092	9234716805
16	6283547091	

As in the case of the square arrangements, the alternative patterns for a single common sum differ by just two transpositions of consecutive numbers, and these two transpositions are separated by a single position. As always, the common sum must be no less than $(5n-3)/2$ and no greater than $(7n-3)/2$, which gives the limits of 11 and 16 for n=5.

For hexagonal arrangements we have n=6, so the only possible common sums are in the range from 13.5 to 19.5, i.e., the set 14, 15, 16, 17, 18, and 19. The essentially distinct solutions for each of these common sums are listed below, along with the corresponding values of V and E, and (in the right-most columns) the sums of the squares of the vertex and edge numbers.

S=14, V=18, E=48

6 3 5 7 2 11 1 9 4 10 0 8	82 424
5 3 6 7 1 11 2 8 4 10 0 9	82 424
8 5 1 11 2 9 3 7 4 10 0 6	94 412

S=15, V=24, E=42

0 6 9 2 4 10 1 11 3 5 7 8	156 350

S=16, V=30, E=36

6 0 10 1 5 7 4 9 3 11 2 8	190 316
7 3 6 8 2 4 10 1 5 11 0 9	214 292
8 5 3 11 2 4 10 0 6 9 1 7	214 292
2 10 4 1 11 5 0 9 7 3 6 8	226 280
0 9 7 5 4 1 11 3 2 8 6 10	226 280
7 9 0 6 10 4 2 11 3 5 8 1	226 280

S=17, V=36, E=30

9 3 5 1 11 2 4 6 7 10 0 8	292 214
11 2 4 8 5 3 9 1 7 10 0 6	292 214

11 2 4 10 3 6 8 0 9 7 1 5	292 214
9 0 8 6 3 4 10 2 5 11 1 7	280 226
9 3 5 8 4 2 11 0 6 10 1 7	280 226
6 4 7 2 8 0 9 3 5 11 1 10	256 250

S=18, V=42, E=24

2 9 7 1 10 0 8 6 4 3 11 5	354 152

S=19, V=48, E=18

11 1 7 2 10 0 9 4 6 8 5 3	412 94
10 0 9 3 7 1 11 2 6 8 5 4	412 94
3 5 11 1 7 4 8 2 9 0 10 6	424 82

These results show some interesting dualities. For example, the first solution with S=14 is the dual of the last solution with S=19, because they have the same split of the numbers into vertex and edge numbers, except that the types are reversed. In other words, the numbers that are assigned to vertices in the first solution are assigned to edges in the last solution, and vice versa. These two solutions are shown pictorially below.

6 3 5	3 5 11
8 7	6 1
0 2	10 7
10 11	0 4
4 9 1	9 2 8

We can generalize the problem by allowing *k* numbers to be assigned to each edge, where *k* is greater than 1. For example, we can construct triangles such as the two shown below

1	1
6 2	6 10
8 9	7 3
4 5 3 7	8 12
	9 2 4 11 5

In general, to construct a triangle with k numbers on each edge (not counting the vertices), let N = k+2 denote how many numbers are on each edge counting the vertices, and then assign the numbers 1, N, and 2N-1 to the three vertices. Then, to fill in the edge numbers, arrange the remaining numbers into three columns, ascending on each row alternately from the left and the right. For example, to construct a triangle with N=6 numbers on each side, we first assign the numbers 1, 6, and 11 to the vertices, and then we arrange the remaining numbers in three columns as shown below

2 3 4
8 7 5
9 10 12
15 14 13

Each column sums to 34, but we actually need columns that sum to 34-5, 34, and 34+5 in order to place these numbers on the edges [6,11], [1,11], and [1,6] of the triangle, respectively. We can do this by leaving the middle column untouched, and making the transpositions (8,5) and (15,13) in the outer columns. This decreases the left-hand column by 5, and increases the right hand column by 5. As an alternate solution, we could make the transpositions (2,4) and (9,12), which would increase the left column by 5 and decrease the right column by 5. Notice that this must always lead to a solution, because the two outer columns will always differ by 2 or 3, so we can form any required adjustment (greater than 1) by transposing pairs.

POLYHEDRA

Suppose we need to schedule 32 golfers into foursomes for five rounds of a tournament, in such a way that no one is in the same foursome with anyone else more than once. This is not too hard, because we have enough players to easily avoid repeated pairings. For example, we can use the simple pattern shown below, where the first round has the 32 players listed horizontally, the second round has them listed vertically, and the subsequent rounds skew the successive columns of the vertical listing by 1, 2, and 3 rows per column, respectively.

Four-some	***Round*** 1	2	3	4	5
1	1 2 3 4	1 9 17 25	1 16 23 30	1 15 21 27	1 14 19 32
2	5 6 7 8	2 10 18 26	2 9 24 31	2 16 22 28	2 15 20 25
3	9 10 11 12	3 11 19 27	3 10 17 32	3 9 23 29	3 16 21 26
4	13 14 15 16	4 12 20 28	4 11 18 25	4 10 24 30	4 9 22 27
5	17 18 19 20	5 13 21 29	5 12 19 26	5 11 17 31	5 10 23 28
6	21 22 23 24	6 14 22 30	6 13 20 27	6 12 18 32	6 11 24 29
7	25 26 27 28	7 15 23 31	7 14 21 28	7 13 19 25	7 12 17 30
8	29 30 31 32	8 16 24 32	8 15 22 29	8 14 20 26	8 13 18 31

However, suppose we have only 24 players, instead of 32. This is a bit more challenging, since we have to work harder to avoid repeated pairings. This is a standard problem in combinatorial design. One simple approach would be to start with four pair-wise orthogonal latin squares of size 5x5. (For any prime power *N* there exist N-1 pairwise orthogonal Latin squares of size NxN.)

0 1 2 3 40 1 2 3 4	0 1 2 3 4	0 1 2 3 4
1 2 3 4 02 3 4 0 1	3 4 0 1 2	4 0 1 2 3
2 3 4 0 14 0 1 2 3	1 2 3 4 0	3 4 0 1 2
3 4 0 1 21 2 3 4 0	4 0 1 2 3	2 3 4 0 1
4 0 1 2 33 4 0 1 2	2 3 4 0 1	1 2 3 4 0

Here we've used the integers (mod 5), and the values of the kth array for *k*=1,2,3,4 are given by

$$A_k(i, j) = k\,(i-1) + (j-1) \pmod 5$$

for *i*,*j* = 1,2,3,4,5. These arrays are called "latin squares" because each element appears exacly once in each row and column. They are said to be "pairwise orthogonal" because if you super-impose one on top of the other you get all 25 distinct combinations (*i*, *j*) with *i*, *j* = 0,1,2,3,4. In other words, each element of one array matches up with each element of the other array exactly once.

So, if we let the elements of the first array represent the first five players, and the elements of the second array represent the next five players, and so on, we can create 25 foursomes from the 20 players by superimposing the four arrays. Since the arrays are pairwise orthogonal, we are assured that no one plays in a foursome with the same person twice.

But we actually need 30 foursomes involving 24 players. This can be done by mixing in four more players, taking advantage of the fact that none of the five players in a given array have yet played with each other. Thus, numbering the players 1 through 24, we have the following five days worth of foursomes. Each of the 24 players plays once per day, and no one plays with the same person twice.

Four-Some	*Day 1*	*Day 2*	*Day 3*	*Day 4*	*Day 5*
1	1 6 11 16	22 8 14 20	3 10 12 19	4 7 15 18	5 9 13 17
2	2 7 12 17	23 9 15 16	4 24 13 20	5 8 24 19	1 10 14 23
3	3 8 13 18	24 10 11 17	5 21 14 16	1 9 21 20	2 6 15 24
4	4 9 14 19	5 6 12 18	1 22 15 17	2 10 22 16	3 7 11 22
5	5 10 15 20	21 7 13 19	2 23 11 18	3 6 23 17	4 8 12 21
6	21 22 23 24	1 2 3 4	6 7 8 9	11 12 13 14	16 18 19 20

This is good, but we can do even better. It's possible to schedule 24 players into foursomes for SEVEN days in such a way that no two players are in the same foursome twice. Obviously this means each player is matched with 21 other players, so he is never matched with 3 of the others. This partitions the players into eight groups of 3, consisting of those who never play together.

For convenience, let's say these exclusion groups are i,i+8,i+16 modulo 24 for i=1 to 8. Our arrangement will be symmetrical in these sets. Now, the players 1, 9, 17 never play together, so we can put them in foursomes #1, #2, and #3 respectively on each of the seven days. Then we can fill in the other 21 players in Foursome #1 in such a way we can create foursomes #2 and #3 simply by adding 8 (modulo 24) to each player of the preceding foursome.

Then we can create Foursome #4 using just the 21 players, excluding players 1, 9, and 17, in such a way that (again) Foursomes #5 and #6 are simply the preceding foursome incrementd by 8 (mod 24). The result is summarized below:

Four-some *Day*	1	2	3	4	5	6
1	1 2 11 21	9 10 19 5	17 18 3 13	4 7 6 24	8 12 14 15	16 20 22 23
2	1 3 12 22	9 11 20 6	17 19 4 14	5 8 7 18	2 13 15 16	10 21 23 24
3	1 4 13 23	9 12 21 7	17 20 5 15	6 2 8 19	3 10 14 16	11 18 22 24
4	1 5 14 24	9 13 22 8	17 21 6 16	7 3 2 20	4 10 11 15	12 18 19 23
5	1 6 15 18	9 14 23 2	17 22 7 10	8 4 3 21	5 11 12 16	13 19 20 24
6	1 7 16 19	9 15 24 3	17 23 8 11	2 5 4 22	6 10 12 13	14 18 20 21
7	1 8 10 20	9 16 18 4	17 24 2 12	3 6 5 23	7 11 13 14	15 19 21 22

For another interesting problem, suppose you're planning a golf tournament in which 12 players will each play one round per day for 4 days. You need to decide how to split up the 12 players into 3 foursomes each day, and you want to choose the foursomes to minimize the number of repeated pairings.

Ideally you'd like to have each pair of players placed together in the same foursome only once, but that's obviously not possible with the conditions of this problem, so the objective is to minimize the repeated pairings AND to make the schedule as symmetrical as possible for the 12 players. What's the "best" (most equally distributed) solution?

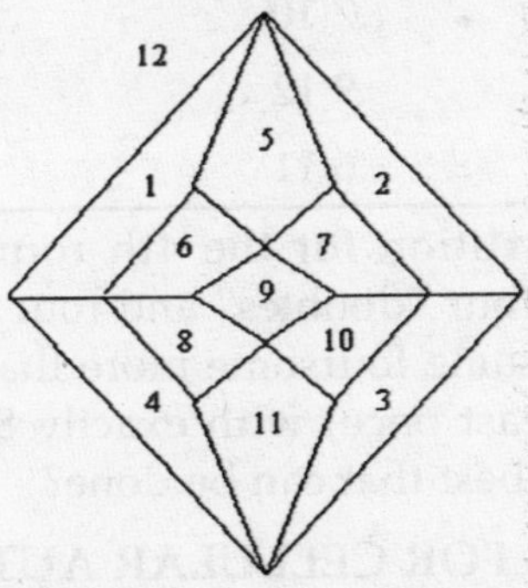

In this case it isn't obvious, a priori, how to even precisely characterize the "best" solution, because minimizing the number of repeated pairs doesn't necessarily give a symmetrical solution. One approach is to think of a fairly symmetrical polyhedron with suitable properties. The rhombic dodecahedron comes to mind, because it has 12 faces, each of which has 4 neighbors, and there are 3 dual pairs of mutually exclusive partitions of the 12 faces into three sets of four. Here's the graph of this solid

The only pairs of faces whose sets of neighbors are mutually exclusive are [(1,10),(2,8)], [(3,6),(4,7)], [(5,11),(9,12)]. The largest number of faces that are all mutually non-neighbors is four, and each face is a member of exactly three such sets. This gives the partitions

1 2 8 10	1 2 9 11	1 3 7 8
3 4 5 9	3 4 6 7	2 4 6 10
6 7 11 12	5 8 10 12	5 9 11 12

To give a 4th partition, notice that "1" and "10" are each already paired twice with "2" and twice with "8", and these are the dual sets of opposite faces [(1,10),(2,8)]. Also, the opposite faces "1" and "10" are already paired once with each other. Similar pairings exist for the other two dual sets. This suggests that we create three new foursomes by combining non-dual pairs. There are eight distinct ways of doing this. For example, if we combine (1,10) with (4,7), and combine (2,8) with (9,12), and so on, we have

1 4	7 10
2 8	9 12
3 5	6 11

Using this partition for the 4th round, each player's schedule includes four "doubles" and four "singles". No two people are on the same foursome more than twice, and each person plays (at least once) with exactly 8 (of the 11) other players. Is this the best that can be done?

STRINGS FOR CELLULAR AUTOMATA

In the note on Cellular Automata we discussed a class of simple linear two state automata in which each field value $f_{t+1}(j)$ is either 0 or 1, and is a fixed function of the three prior values $f_t(j-1)$, $f_t(j)$, and $f_t(j+1)$. We saw how the 256 distinct functions ("rules") of this type can be placed into 88 equivalence classes up to direction reversal and field value complementation. Furthermore, we noted that for any given rule there may be impossible (excluded) strings, i.e., strings that are not the successor of any string. For any given rule, we can list the total number of k-bit excluded strings for $k = 1, 2,...$ etc. This gives sequences of integers, and we found that there are only 31 distinct sequences. The members of each of the 88 equivalence classes of rules fall into one of these 31 "exclusion classes". The largest single exclusion class consists of the rules that do not exclude any strings. At the other extreme, Rule 0 excludes all but one string. In between these two extremes are the other 29 exclusion classes, with varying degrees of exclusivity.

Naturally any string that contains an excluded string is also excluded, so we are led to focus on the *primitive* excluded strings, i.e., strings that are excluded but that do not contain any proper substring that is excluded.

Typically the number of primitive excluded strings is a partition function. For example, as discussed in Cellular Automata, the primitive excluded strings for exclusion class 62 (which includes Rule 110) are of the form "01*10" where the "*" consists of a certain number of "0" bits followed by some number (possibly zero) of "111" strings each followed by one or more "0" bits.

Consequently, letting $P_{62}(k)$ denote the number of primitive excluded k-bit strings for Class 62, we find that this is equal to the number of ordered partitions of k-4-3n into exactly n+1 positive integers, for all *n* such that k-4-3n is greater than or equal to n+1. Thus the values of *n* range from 0 to the greatest integer less than or equal to (k-5)/4. The values of $P_{62}(k)$ for $k = 1$ to 17 are listed below.

0, 0, 0, 0, 1, 1, 1, 1, 2, 3, 4, 5, 7, 10, 14, 19, 26,...

These values satisfy the 4th-order linear recurrence relation

$$P_{62}(k) = P(k-1) + P(k-4)$$

where we have omitted the subscript "62" on the right-hand *P* terms for convenience. The characteristic polynomial x^4-x^3-1 is irreducible, and has the approximate roots

1.380277, -0.819172, 0.219447 ± 0.914473i

Therefore, the sequence asymptotically increases by a

factor of about 1.38 as k is incremented. Of course, since any string that contains one of these primitive excluded strings is excluded, the total number of excluded k-bit strings for Class 62 increases more rapidly.

The values of this number, which we will denote as $E_{62}(k)$, are listed below for $k = 1$ to 17.

0, 0, 0, 0, 1, 5, 16, 43, 107, 255, 589, 1328, 2940, 6419, 13862, 29668, 63024,...

These values, too, satisfy a linear recurrence, namely the 6th-order recurrence

$$E_{62}(k) = 5E(k-1) - 9E(k-2) + 8E(k-3) - 4E(k-4)$$
$$-E(k-5) + 2E(k-6)$$

(Again we have omitted the subscript "62" on the right-hand side.) The characteristic polynomial of this recurrence factors

$$x^6 - 5x^5 + 9x^4 - 8x^3 + 4x^2 + x - 2 = (x-1)(x-2)$$
$$(x^4 - 2x^3 + x^2 - x - 1)$$

The total number of k-bit strings is obviously 2^k, so the fraction of k–bit strings that are excluded is $E(k)/2^k$, which approaches 1 as k increases. This isn't surprising, because all it takes is one small primitive excluded string to exclude "almost all" sufficiently long strings. In fact, several of the exclusion classes (0, 3, 4, 7, 14, and 29) have only a single primitive excluded string.

In general, for rule classes that have easily-characterized primitive excluded strings, it seems to be possible to find a linear recurrence for the P and E sequences. Here is a short list of characteristic polynomials and their factorizations for several E-classes:

$$E_0 : x^2 - 3x + 2 = (x-2)(x-1)$$

$$E_2 : x^4 - 3x^3 + 2x^2 - x + 2 = (x-2)(x^3 - x^2 - 1)$$

$$E_3 : x^4 - 4x^3 + 5x^2 - 3x + 2 = (x-2)(x^3 - 2x^2 + x - 1)$$

$$E_4 : x^3 - x^2 + x + 2 = (x-2)(x^2 - x - 1)$$

$$E_7 : x^5 - 4x^4 + 4x^3 + x^2 - 3x + 2 = (x-2)(x^4 - 2x^3 + x - 1)$$

$$E_{14} : x^4 - 3x^3 + x^2 + x + 2 = (x-2)(x^3 - x^2 - x - 1)$$

$$E_{18} : x^6 - 5x^5 + 8x^4 - 4x^3 + x - 2 = (x-2)(x^2 - x - 1)$$
$$(x^3 - 2x^2 + x - 1)$$

$$E_{25} : x^6 - 4x^5 + 3x^4 + 4x^3 - 4x^2 + x - 2 = (x-2)$$
$$(x^5 - 2x^4 - x^3 + 2x^2 - 1)$$

$$E_{29} : x^5 - 4x^4 + 4x^3 + x - 2 = (x-2)(x-1)(x^3 - x^2 - x - 1)$$

$$E_{46} : x^4 - 4x^3 + 4x^2 + x - 2 = (x-2)(x-1)(x^2 - x - 1)$$

$$E_{62} : x^6 - 5x^5 + 9x^4 - 8x^3 + 4x^2 + x - 2 = (x-2)(x-1)$$
$$(x^4 - x^3 + x^2 - x - 1)$$

Not surprisingly, each of the characteristic polynomials contains a factor of $(x-2)$, and therefore 2 is the dominant root, causing each sequence (asymptotically) to roughly double on each step. In addition, we see that several other factors are shared by two or more of these polynomials. The factor $(x-1)$ signifies a constant offset in the sequences for classes 0, 29, 46, and 62. We also see the "Fibonacci factor" (x^2-x-1) in the polynomials for classes 4, 18, and 46, the 3rd-order Fibonacci factor (x^3-x^2-x-1) for classes 14 and 29, and the factor (x^3-2x^2+x-1) for classes 3 and 18.

We can also find recurrences for the numbers of primitive excluded strings for some of the rule classes. Here is a short list of the characteristic polynomials and their factorizations for a few of the rule classes with non-trivial P(k) sequences.

$$P_{18} : x^4 - 2x^3 + x^2 - 1 = (x^2 - x - 1)(x^2 - x + 1)$$

$$P_{25} : x^4 - x^2 - x - 1 = (x+1)(x^3 - x^2 - 1)$$

$$P_{62} : x^4 - x^3 - 1 = \text{irreducible}$$

As in the case of Class 62, the primitive excluded strings for Classes 18 and 25 can be expressed in terms of partitions. For Class 18 the primitive excluded strings are "111" plus all the strings of the form 11*11 where "*" signifies an odd number of individual "1" bits separated by at least one "0" from every other "1". Thus the value of $P_{18}(k)$ is the number of ordered partitions of k–4–n into exactly n + 1 positive parts for odd integers n such that k–4–n is greater than or equal to n +1.

Hence n ranges over the odd integers from 1 to the greatest odd integer less than or equal to (k-5)/2. To illustrate, with k=16 the permissible values of n are 1, 3, and 5, so we need to count the numbers of ordered partitions of 11 into two parts, of 9 into four parts, and of 7 into six parts. This gives 10, 56, and 6 respectively, so the total is $P_{18}(16) = 72$. Similarly the primitive excluded strings for Class 25 can be expressed in terms of restricted partitions, although the restriction is slightly more complicated.

The remaining six non-trivial P(k) seauences (i.e., those for the classes 37, 22, 94, 73, 41, and 26) do not appear to satisfy a linear recurrence of order 8 or less. It is not known whether they satisfy any higher-order linear recurrence, or if they have simple generating functions.

HAMILTON CYCLES ON MCCAULEY GRAPHS

Suppose we wish to arrange a string of 8 bits in a circle such that every possible 3-bit string occurs exactly once. One possible solution is 00011101, which (with wrap-around) gives the eight sequences of 3 bits

000	001	011	111	111	110	101	010	100
0	1	3	3	7	6	5	2	4

Each of the eight numbers in this sequence must be given by shifting the preceding number one bit to the left (i.e., multiplying by 2), adding either 0 or 1, and truncating the most significant bit (i.e., taking the residue modulo 8). Thus, beginning with $x = 0$, each succeeding number in the sequence is given by either $2x$ (mod 8) or $2x + 1$ (mod 8). We can express the above solution just in terms of the sequence of eight residues 0, 1, 3, 7, 6, 5, 2, 4. There is only one other solution (up to rotations), namely, 0, 1, 2, 5, 3, 7, 6, 4, corresponding to the string 00010111. These two solutions are just the reflections of each other (up to rotation). Obviously the sequence of bits represents the summand b in the operation $2x + b$ (mod 8) at each step, and this is also the sequence of odd/even numbers in the result.

As noted by Joe McCauley, we can find analogous sequences modulo any even number n (not necessarily a power of 2), such that every residue modulo n appears exactly once, and each number is either 2x (mod n) or 2x + 1 (mod n) relative to the preceding number x. Of course, if n is not a power of 2, the evaluations modulo n will not be pure truncations, so these solutions don't correspond exactly to strings of binary bits taken k bits at a time as they do when n equals 2^k. Thus a more general representation is in terms of a directed graph consisting of n vertices numbered 0 to n-1, with a directed edge from each vertex j to the two vertices $2j$ and $2j + 1$ (mod n). We will call this a McCauley graph, and consider the Hamilton cycles of these graphs. There are no Hamilton cycles for McCauley graphs of odd degree, so we need consider only the graphs for even values of n. The graphs for $n = 2, 4, 6, 8$, and 10 are shown below.

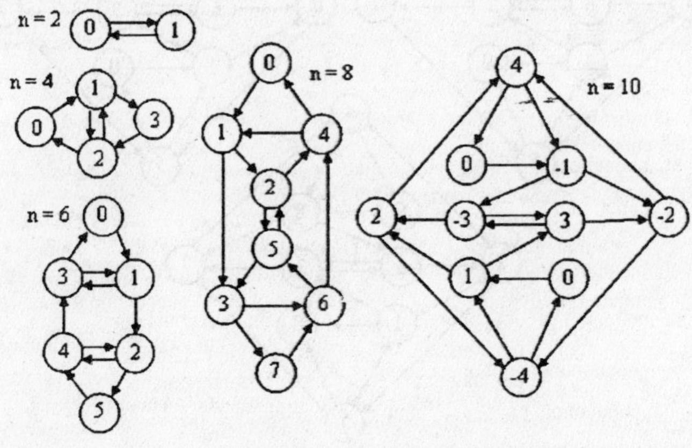

Each of these is a planar graph, meaning that it can be drawn on a two-dimensional surface with no crossing lines. (For clarity we haven't shown the self-edges for the "0" and "n/2" vertices.) It's easy to see that there is only one Hamilton cycle for each of the cases $n = 2$, 4, and 6. For the McCauley graph of degree $n = 8$ there are two distinct Hamilton cycles, as shown in the figures below.

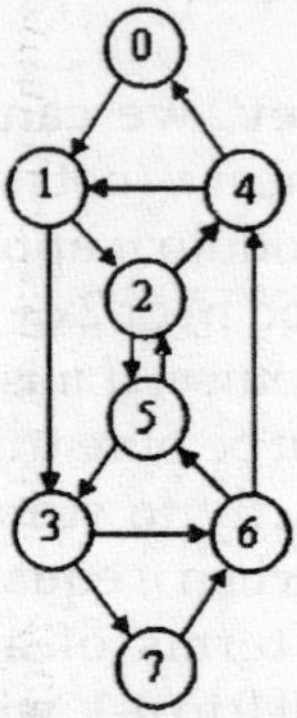

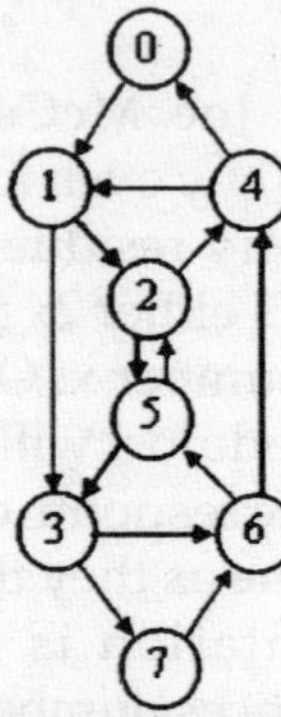

For the McCauley graph of degree $n = 10$ there are three distinct Hamilton cycles, as shown in the figures below.

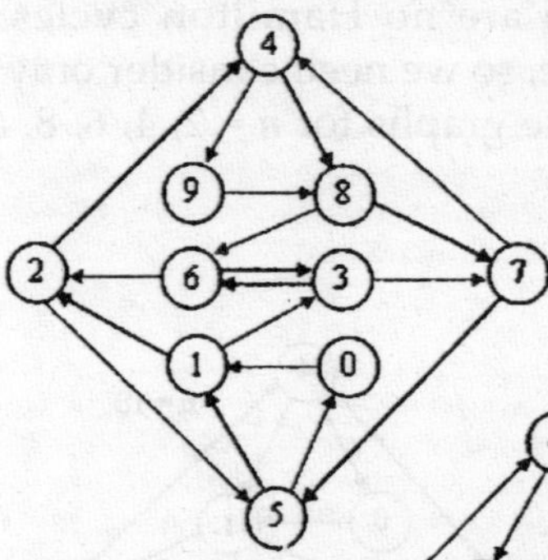

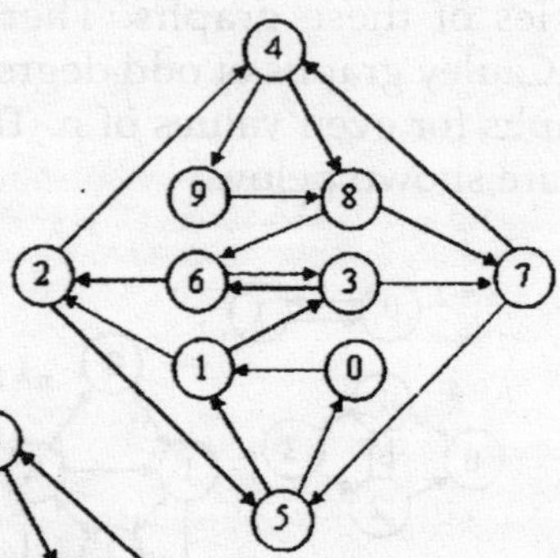

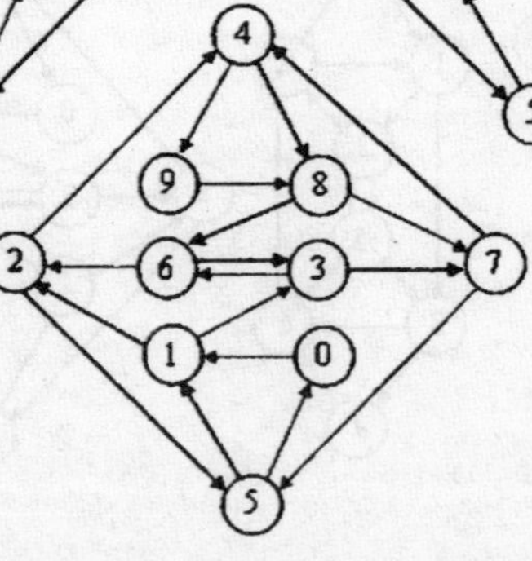

We also note that the graphs are symmetrical if we reduce the residues to the smallest signed residues modulo n–1. For the case $n = 14$ there are seven distinct Hamilton cycles, represented by the sequences of vertices listed below.

0	1	2	4	8	3	6	13	12	11	9	5	10	7
0	1	2	4	9	5	10	6	13	12	11	8	3	7
0	1	2	4	9	5	11	8	6	6	13	12	10	7
0	1	2	5	10	6	13	12	9	9	4	8	3	7
0	1	2	5	11	9	4	8	6	6	13	12	10	7
0	1	3	6	13	12	12	8	4	4	9	5	10	7
0	1	3	6	13	12	12	9	8	8	2	5	10	7

Reducing these to the smallest signed residues modulo 13 shows the symmetries more clearly.

0	1	2	4	–5	3	6	3	6	0	–1	5	–3	–6
0	1	2	4	–4	5	–3	5	–3	6	0	–5	3	–6
0	1	2	4	–4	5	–2	5	–2	–5	3	–1	–3	–6
0	1	2	5	–3	6	0	6	0	–1	–2	–5	3	–6
0	1	2	5	–2	–4	4	–4	4	–5	3	–1	–3	–6
0	1	3	6	0	–1	–2	–1	–2	–5	2	5	–3	–6
0	1	3	6	0	–1	–2	–1	–2	–4	4	5	–3	–6

McCauley noted that the numbers of Hamilton cycles alternate between odd and even, i.e., the number is odd if and only if n/2 is odd. He asked if there are any other patterns in these numbers, and if there is any other way of producing them. It turns out that there is a nice "closed form" number-theoretic expression for these numbers.

First we make a few more observations on the table. Letting H(n) denote the number of cycles of size n, we note that $H(2(2n)) = 2^{n-1}H(2n)$ for all n. For example, H(36) = (256)H(18). This takes care of H(n) for all n divisible by 4, so we need only determine the values of H(n) for n = 2k where k is odd. Now we observe that if $n = 2p$ for a prime p we usually have

$$H(2p) = \frac{2^{p-1}}{p} \quad \text{for } p = 3,5,11,13,19,29$$

Of course, by Fermat's theorem, we're assured this quantity is an integer for any prime p. For example, H(2‴11)

$= (2^{10} - 1)/11 = 93$. However, this doesn't cover the primes 7, 17, 23, or 31. These are precisely the primes (in the range of the table) for which H(2*p*) is divisible by *p* itself. Moreover, for each of these four primes the value of H(2p) is of the form pm^2 for some integer m. In other words, H(2p)/p is a square. Furthermore, in three of these cases (*p* = 7, 17, and 23) we find that $H(2p) = pm^2 = (2p-2)/(2p)-2m$, so we can solve for $m = (2^{(p-1)/2}-1)/p$, which implies

$$H(2p) = \frac{\left(2^{(p-1)/2} - 1\right)^2}{p} \quad \text{for} \quad p = 7,17,23$$

The only remaining prime case (in the table) is *p* = 31, for which we have $h(2'''31) = 31^5$. It's also clear from the table that the values of *h*(2*n*) for odd composite *n* are related to divisors of 2^q-1 for primes *q* that divide *n*.

To summarize these observations, let h(n) denote H(2n). We don't lose anything this way, since there are no cycles with odd lengths. With this convention, the observations mentioned above can be summarized as follows: $h(2n) = 2^{n-1}h(n)$

for all positive integers n $h(p) = \dfrac{2^{\phi(p)} - 1}{p}$

for primes *p* = 3,5, 11, 13, 19, 29 $h(p) = \dfrac{\left(2^{\phi(p)/2} - 1\right)^2}{p}$

for primes *p* = 7, 17, 23

where *f*(*n*) is Euler's totient function. The prime 31 didn't fit into these patterns, but notice that we

have $h(p) = \dfrac{\left(2^{\phi(p)/6} - 1\right)^6}{p}$

for primes p = 31,....

so we are led to suspect that *h*(*p*) for every prime *p* is of the form

$$h(p) = \frac{\left(2^{\phi(p)/k} - 1\right)^k}{p}$$

for some divisor k of $f(p)$. Furthermore, for each *power* of a prime, we apparently have

$$h(p^r) = \frac{1}{p^r} \prod_{j=1}^{r} \left(2^{\phi(p^j)/k} - 1\right)^k$$

for a suitable divisor k of $f(p)$. Also, notice that for at least some odd primes p,q we have

$$h(pq) = h(p)h(q)\left(2^{\phi(pq)/k} - 1\right)^k$$

where k again is a suitable divisor of $f(n)$. To this point we have left the value of "k" unspecified, other than to say it is a divisor of f, but notice that for each integer N the suitable value of $\phi(N)/k$ is equal to the *order* of 2 modulo N. Letting ord(m,N) denote the least exponent s such that $m^s - 1$ is divisible by N, we have $k = \phi(N)/\text{ord}(2,N)$.
This is consistent with the basic operation, which involves cycles generated by iterating $x \to 2x$ (or $2x+1$) modulo N. Combining all these relations, we conjecture that, for any integer $n = 2^r m$ where m is odd we have

$$h(n) = H(2n) = \frac{2^{n-m}}{n} \prod_{d/m} \left(2^{\phi(2,d)} - 1\right)^{\phi(d)/\text{ord}(2,d)}$$

Thus we arrive at an explicit formula for the number of distinct Hamilton cycles of McCauley graphs of degree n. (Note that 2nH(2n) could be regarded as the actual number of Hamilton cycles of degree $2n$, because H(n) was restricted to those beginning with 0, each of which represents 2n others by rotation.)

As an example, suppose we wish to determine the value of H(42), which equals h(21). In this case $n = m = 21$, which has the divisors 3, 7, and 21. Using the values $\phi(3) = 2$, $\phi(7) = 6$, $\phi(21) = 12$, and ord(2,3) = 2, ord(2,7) = 3, ord(2,21) = 6, the preceding formula gives

$$h(21) = \frac{1}{21}\left(2^2 - 1\right)\left(2^3 - 1\right)^2\left(2^6 - 1\right)^2 = 27783$$

in agreement with the table value of H(42).

n = 10

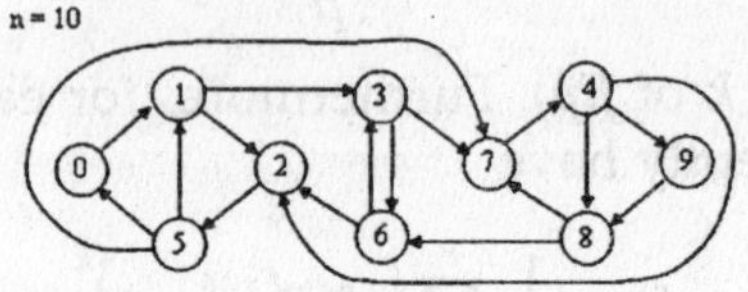

n = 12

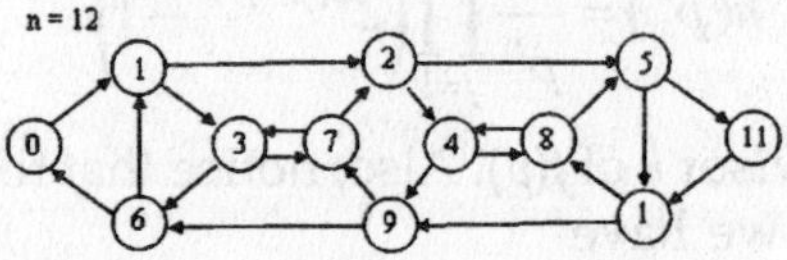

n = 14

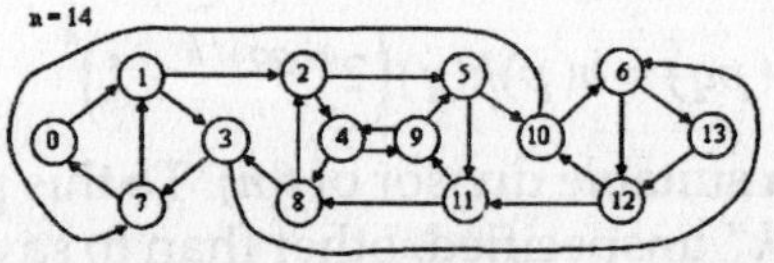

Notice that $h(n)$ is not a simple multiplicative function, because of the cross-product terms involving $2^{\mathrm{ord}(pq)} - 1$. Still, it's an interesting definition based on the divisors. (Can this can be inverted using something like the Mobius function?) This formula is valid for all the values in the table, which is fairly persuasive, but I haven't proven that it gives $h(n)$ for all n. Still, this is an interesting example of how much structure can be inferred (or at least guessed) about a combinatorial function strictly from examining its values, without even considering its definition.

The McCauley graphs are all planar up to $n = 14$, as shown by the figures below. The first non-planar McCauley graph is for $n = 16$, shown in the figure below.

n = 16

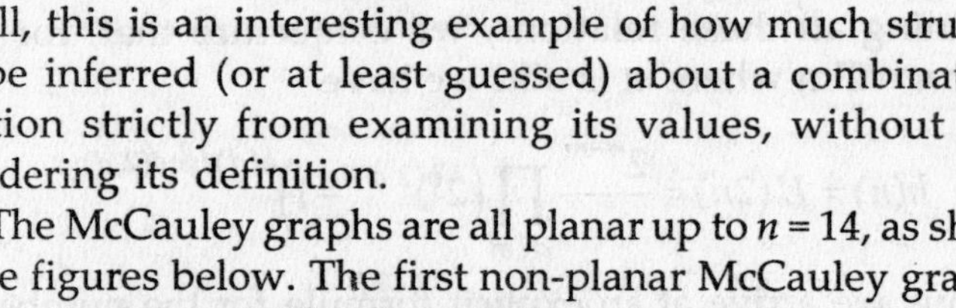

Incidentally, the sequence of integers h(n) is also the number (up to rotation) of $n \times n$ circulant invertible matrices over the field of integers modulo 2. A circulant matrix is one whose rows are all just rotated versions of a single row. For example, the following is a 4 *x* 4 circulant matrix:

$$\begin{bmatrix} 1 & 2 & 3 & 4 \\ 4 & 1 & 2 & 3 \\ 3 & 4 & 1 & 2 \\ 2 & 3 & 4 & 1 \end{bmatrix}$$

A matrix is invertible if its determinant is non-zero. So, an invertible circulant matrix over the integers modulo 2 is a circulant matrix whose elements are either 0 or 1 and whose determinant is odd (i.e., congruent to 1 modulo 2). Since each row is just a rotated version of one of the rows, we can represent each such matrix (or rather the equivalence class of matrices up to rotation) by just the top row. The matrices for the first several values of n are indicated below.

$n=2$	$n=3$	$n=4$	$n=5$	$n=6$	$n=7$
10	100	1000	10000	100000	1000000
		1110	11100	110100	111000
			11010	110010	1100100
				111110	1010100
					1111100
					1111010
					1110110

We can generalize the above discussion to other moduli. For example, instead of considering all cycles of integers of length n such that each term s_{k+1} is either $2s_k$ or $2s_k + 1$ modulo n, we could allow each term to be of the form $3s_k$, $3s_k + 1$, or $3s_k + 2$ modulo n. For example, the four solutions (up to rotation) of length 8 are

0 1 4 6 3 2 7 5

0 2 7 6 3 1 4 5

0 1 3 2 7 6 4 5

0 1 4 5 7 6 3 2

Obviously each of these cycles is uniquely characterized by the sequence of "remainders" modulo 3, so they each correspond to an integer with those digits in base 3. The numbers of cycles for $n = 2$ to 15 are listed below:

n	cycles	n	cycles
2	1	9	14
3	2	10	12
4	2	11	64
5	2	12	64
6	4	13	42
7	4	14	44
8	4	15	512

Recall that with base 2 there were no cycles except when n is divisible by 2. With base 3 there are clearly more cycles when n is divisible by 3 than when it is not, but there do exist cycles in the latter case.

PYTHAGOREAN GRAPHS

Is it possible to fill an infinite square array with distinct integers such that the sum of the squares of any two adjacent numbers is a square? To illustrate, the following is a 4x4 array with the desired property

1836	105	252	735
1248	100	240	700
936	75	180	525
273	560	1344	3920

The key to this problem is to recognize that the row and column conditions are independent. Thus, for any positive integer N we can easily construct a square array of size (N+1)x(N+1) consisting of the values

$$A[m,n] = (12/5)^{m} (24/7)^{n} 35^{N} = 2^{(2m+3n)} 3^{(m+n)} 5^{(N-m)} 7^{(N-n)}$$

for $m, n = 0,1,2,..,N$. These values are all integers and no two of them have the same number of factors of 5 and 7, so they are all distinct. Although this gives a recipe for arbitrarily large arrays, it doesn't give an *infinite* array. R. Mentock provided the following nice construction of an infinite array.

For $m = 3M + a$, $n = 3N + b$, define F(m, n) by

$F(m, n) = 6^M * 132^N * f(a, b)$

where

$f(0,0) = 6 * 7$	$f(0,1) = 6 * 24$	$f(0,2) = 6 * 143$
$f(1,0) = 8 * 7$	$f(1,1) = 8 * 24$	$f(1,2) = 8 * 143$
$f(2,0) = 15 * 7$	$f(2,1) = 15 * 24$	$f(2,2) = 15 * 143$

Another, shorter, solution: For $m = 2M + a$, $n = 2N + b$, define

F(m, n) by

$F(m, n) = 10^M * 76^N * f(a, b)$

where

$f(0,0) = 7 * 17$	$f(0,1) = 7 * 144$
$f(1,0) = 24 * 17$	$f(1,1) = 24 * 144$

Both of the above "infinite" solutions actually fill only one quadrant of the plane. To completely "tile" the plane with Pythaogrean legs, Mentock notes that

Each of the two solutions are actually combinations of two sequences, each defined along the axes. The other points are just products of these, i.e., G(m, n) = G(m) * G(n). Since we have four of them, we start at zero and assign values to the four semi-axes. G(0, 0) is the LCM of all G(0), and that number is propogated through. Then G(m,n) = G(m) * G(n) like above, *and there we go.* A slightly different approach was suggested by Dave Radcliffe, who wrote

Given $F(m, n) = 10^M * 76^N * f(a,b)$ where

$f(0,0) = 7 * 17$	$f(0,1) = 7 * 144$
$f(1,0) = 24 * 17$	$f(1,1) = 24 * 144$

For $m, n >= 0$ define

$G(m, n) = F(m, n) * 11 * 13$

$G(-m-1, n) = F(m, n) * 11 * 84$

$G(-m-1,-n-1) = F(m, n) * 60 * 84$

$G(m, -n-1) = F(m, n) * 60 * 13$

Since no F(m,n) is divisible by 11 or 13, G is also one-to-one.

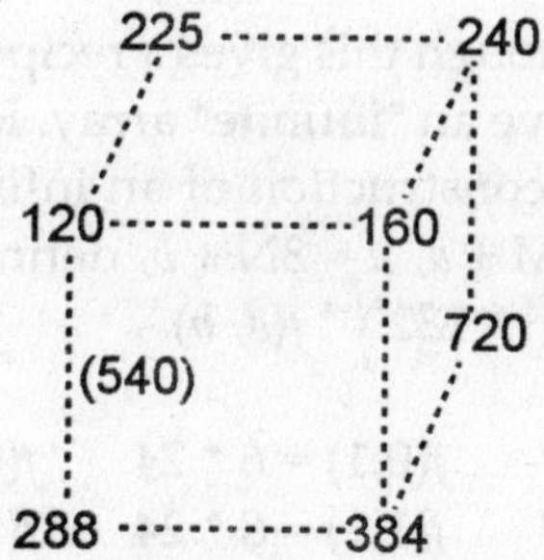

Here the values increase by factors of 4/3, 12/5, and 15/8 in the three principle directions. (Presumably it's possible to fill an infinite 3D orthogonal lattice with distinct numbers in this way.)

This leaves only the dodecahedron. It turns out that it is possible to populate the vertices of a dodecahedron with distinct numbers such that the sum of the squares of each pair of adjacent numbers is a square.

a = 155040	f = 649800	k = 802560	p = 2006400
b = 290700	g = 1438965	l = 601920	q = 180576
c = 897600	h = 3762000	m = 451440	r = 689700
d = 1683000	i = 2613600	n = 240768	s = 846450
e = 177650	j = 1504800	o = 1070080	t = 275880

Another related question, raised by Dave Radcliffe, is whether it possible to connect any two positive integers (greater than 4) by a series of Pythagorean legs. (As Dave put it, "consider a graph G with a vertex for each integer $n > [4]$, and an edge joining m and n if $m^2 + n^2$ is a square. Is this graph connected?") Notice that the numbers 3,4 are a separate graph and don't connect to anything else. But for $n > 4$ it certainly appears to be true that the graph is connected.

If you look at this as a special case of the more general "connection criterion" $m^2 + n^2 = k$*square (where k is a square-free integer expressible as a sum of two squares), it seems the case k=1 is unique in giving a connected graph for all the integers (above some number). In contrast, consider the graph based on $m^2 + n^2 = 2$*square. In this case the naturals split into mutually disjoint graphs, each of which is a multiple of the basic connected graph consisting of the numbers

1, 7, 17, 23, 31, 41, 47, 49, 71, 73, 79, 89, 97, 103, 113, 119,...

These are exactly the numbers whose prime factors are all congruent to +1 or -1 modulo 8. In other words, they are the primes p such that 2 is a square (mod p). The graph based on $m^2 + n^2 = 5$*square splits into mutually disjoint graphs, each of which is a multiple of

1, 2, 4, 8, 11, 16, 19, 22, 29, 31, 32, 38, 41, 44, 58, 59, 61, 62,...

These appear to be exactly the numbers expressible as products of the primes 2, 11, 19, 31, 41, 59, 61, 71, 79, 89, 101,... which I think are the primes p such that 5 is a square (mod p). The graph based on $m^2 + n^2 = 10$*square splits into disjoint graphs that are each multiples of

1, 3, 9, 13, 27, 31, 37, 39, 41, 43, 53, 67, 71, 79, 81, 83,...

Again it appears there are all the products of a particular set of primes, in this case the primes 3, 13, 31, 37, 41, etc., and these look like the primes modulo which 10 is a square.

So, we might be tempted to conclude that the graph of the naturals with the connection criterion $m^2 + n^2 = k$*square splits into disjoint graphs, each of which is a multiple of a graph consisting of the numbers expressible as a product of primes p not dividing k and such that k is a quadratic residue mod p. (This explains why the case k=1 gives just a single totally connected graph.) However, the case k=13 seems to behave differently. It's basic connected set consists of

1, 8, 12, 18, 27, 34, 51, 53, 79, 86, 92, 96, 103, 116, 122,...

The set of primes for this case should be 2,3,13,17,23,29,43,etc. but there seems to be something else going on here.

ELECTORAL DRAWS

The principle of hereditary kingship became firmly established in most countries of western Europe during the middle ages, due partly to the historical accident that, for several centuries, each succeeding king by the time of his death had an adult son who seemed at least nominally capable of becoming the new king. Germany, however, was an exception. During these same centuries, a series of kings of Germany (who also held the title of Holy Roman Emperor) failed to leave any clear heir, either because they had no sons, or because their sons were still children when the king died. As a result, the question of succession had to be decided by other means. Gradually a system relying on a group of Electors was established. The Electors originally consisted of seven German dignitaries, who met after the passing of each king to select his successor.

The Electors chose the German kings for over 500 years, from the 13th century until they were finally abolished in 1806 during the Napoleonic era. Ironically, just 18 years before the final abolition of the German Electors, a new system of Electors was established for selecting the leader of the United States of America. When the constitution of the United States was written in 1788, the power to select a president was given to a group of Electors. However, unlike the German Electors, these Electors were not hereditary, but were to be themselves elected by the legislative bodies of each state of the union. Of course, the members of the legislative bodies were themselves elected, by popular vote of all the citizens. The framers might have considered making citizenship an elective office as well, but decided instead to let citizenship be hereditary. (It's not so easy to completely eliminate the principle of heredity.) New citizens could also be created by naturalization.

In practice the states soon began to allow the presidential Electors to be chosen directly by popular vote, bypassing the state legislatures.

The original intent of the framers of the constitution was for the presidential Electors to exercise their own judgment in making their selections for President and Vice President, but the rise of the party system led to the practice of having each candidate Elector public ally declare his allegiance to one party, so by voting for a particular Elector the citizen is essentially voting for the presidential and vice presidential candidates of that party. Another innovation came about when the states realized that they could increase their relative influence by adopting a "winner-take-all" policy, whereby *all* the electoral candidates of the party that gets the most popular votes (in that state) are chosen.

By winning the popular vote in a winner-take-all state with nine Electors, the presidential candidate receives all nine electoral votes, whereas if the Electors were chosen in proportion to the popular vote, the presidential candidate would usually get only five to his opponent's four, giving only a one-vote difference. In the belief that candidates will pay more attention to a "winner-take-all" state than to a proportionate state, nearly all the states quickly adopted this policy. (One or two states still allow their Electors to be split between the parties, in proportion to the popular votes, but this is rare.) Of course, once all the states have adopted the winner-take-all policy, the relative advantage of this policy disappears, leaving only the disadvantage that it allows for the possibility of a presidential candidate winning the overall popular votes but losing the electoral vote. This has actually happened at least twice in United States history.

55 Califormnia	10 Arizona	5 Nevada
34 Texas	10 Maryland	5 New Mexico
31 New Yokr	10 Minnesota	5 Utah
27 Florida	10 Wisconsin	5 West Virginia
21 Illinois	9 Alabama	4 Hawaii
21 Pennsylvania	9 Colorado	4 Idaho
20 Ohio	9 Louisiana	4 Maine
17 Michigan	8 Kentucky	4 New Hampshire
15 Georgia	8 South Carolina	4 Rhode Island
15 New Jersey	7 Connecticut	3 Alaska
15 North Carolina	7 Iowa	3 Delaware
13 Virginia	7 Oklahoma	3 D.C.
12 Massachusetts	7 Oregon	3 Montana
11 Indiana	6 Arkansas	3 North Dakota
11 Missouri	6 Kanasas	3 South Dakota
11 Tennessee	6 Mississippi	3 Vermont
11 Washington	5 Nebraska	3 Wyoming

In presidential elections the fifty states–plus the District of Columbia–have a combined total of 538 Electors. The number of Electors from a state equals the number of the states' representatives in the House and the Senate, which depends (for the House) on the population of the state. Based on the population census of 2000, the number of Electors for each of the states are listed below (applicable to the presidential elections of 2004 and 2008):

If we assume every state has a winner-take-all policy, a candidate can either win or fail to win each of the fifty-one regions, so the number of distinct possible outcomes (win patterns) is

$$2^{51} = 2251799813685248 \approx (2.25)10^{15}$$

This represents over two-thousand trillion outcomes.

One possible result of a two-candidate election is an electoral draw, with each candidate receiving exactly 269 electoral votes. This raises the question: In how many distinct ways can a candidate win exactly 269 votes? If we don't distinguish between states with the same number of votes, then we are just dealing with the following multi-set of summands:

3 (8)	4 (5)	5 (5)	6 (5)	7 (4)
8 (2)	9 (3)	10 (4)	11(4)	12 (1)
13 (1)	15 (3)	17 (1)	20 (1)	21 (2)
27 (1)	31 (1)	34 (1)	55 (1)	

A candidate can win 0, 1, ..., or 8 of the three-vote states, and he can win 0, 1, ..., or 5 of the four-vote states, and so on. Thus the total number of distinct win patterns for a candidate (not distinguishing between states with the same number of votes) is

$$(9)(6)^2(5)^3(4)^3(3)^2(2)^8 = 597196800$$

$$\left(1+x^{55}\right)\left(1+x^{34}\right)\left(1+x^{31}\right)\left(1+x^{27}\right)\left(1+x^{21}+x^{2.21}\right)\ldots$$
$$\left(1+x^{3}+x^{2.3}+x^{3.3}+x^{4.3}+x^{5.3}+x^{6.3}+x^{7.3}+x^{8.3}\right)$$

Of these, the number of win patterns giving exactly 269 votes is 40972162 (almost 41 million), which is about one out of 145. This is the coefficient of x^{269} in the expansion of the product Of all the possible vote totals, from 0 to 538, the one with the most ways of occurring is 269.

If we distinguish between the states with the same number of electoral votes, then the number of distinct outcomes is increased, because each of the roughly 41 million win patterns can occur in multiple distinct ways. For example, a given win pattern may contain 5 of the 8 three-vote states, but there are C(8,5) = 56 ways of choosing five of those eight states. Applying the appropriate factors to each win pattern in the set of 269-winning patterns, we find that the number of distinct ways of winning exactly 269 electoral votes, is

1705744125652

which is just over 17 trillion, about 1/132 of all 2^{51} possible outcomes.

Chapter 10

Concept of Binary Relation

OVERVIEW

The relation we are going to study here is an abstraction of relations we see in our everyday life such as those between parent and child, between car and owner, among name, social security number, address and telephone number etc. We are going to focus our attention on one key property which all the everyday relations have in common, define everything that has that property as a relation, and study properties of those relations.

One of the places where relation in that sense is used is data base management systems. Along with hierarchical and network models of data, the *relational model* is widely used to represent data in a database. In this model the data in a database are represented as a collection of relations. Informally, each relation is like a table or a simple file. For example, consider the following table.

Employee

Name	*Address*	*Home Phone*
Amy Angels	35 Mediterranean Av.	224–1357
Barbara Braves	221 Atlantic Av.	301–1734
Charles Cubs	312 Baltic Av.	223–9876

Each row of this table represents a collection of data values such as name, address, and telephone number of a person. Each row is considered an instance of a relation and the table as the collection of the rows is considered a relation, which is the relation we are going to be studying in this chapter.

Operations such as inserting or deleting entries to or from a table, merging two table, finding the intersection of two tables, and searching for certain entries can be described simply and precisely as operations on relations and known mathematical results on relations can be utilized without reinventing them.

The relational model is flexible (easy to expand, easy to modify) and interface to query languages is simple. It is thus widely used today.

DEFINITION

Here we are going to define relation formally, first binary relation, then general n–ary relation. A relation in everyday life shows an association of objects of a set with objects of other sets (or the same set) such as John owns a red Mustang, Jim has a green Miata etc. The essence of relation is these associations. A collection of these individual associations is a relation, such as the ownership relation between peoples and automobiles. To represent these individual associations, a set of "related" objects, such as John and a red Mustang, can be used. However, simple sets such as {John, a red Mustang} are not sufficient here.

The order of the objects must also be taken into account, because John owns a red Mustang but the red Mustang does not own John, and simple sets do not deal with orders. Thus sets with an order on its members are needed to describe a relation. Here the concept of ordered pair and, more generally, that of ordered n–tuple are going to be defined first. A relation is then defined as a set of ordered pairs or ordered n–tuples.

Ordered pair: An ordered pair is a set of a pair of objects with an order associated with them.

If objects are represented by *x* and *y*, then we write an ordered pair as <*x*, *y*> or <*y*, *x*>. In general <*x*, *y*> is different from <*y*, *x*>.

Equality of ordered pairs: Two ordered pairs <*a*, *b*> and <*c*, *d*> are equal if and only if $a = c$ and $b = d$.

For example, if the ordered pair <*a*, *b*> is equal to <1, 2>, then $a = 1$, and $b = 2$. <1, 2> is not equal to the ordered pair <2, 1>.

Binary relation: A binary relation from a set *A* to a set *B* is a set of ordered pairs *<a, b>* where *a* is an element of *A* and *b* is an element of *B*.

When an ordered pair *<a, b>* is in a relation *R*, we write *a R b*, or *<a, b>* $\in$ *R*. It means that element *a* is related to element *b* in relation *R*.

When *A* = *B*, we call a relation from *A* to *B* a (binary) relation on *A*.

Cartesian product: The set of all ordered pairs *<a, b>*, where *a* is an element of *A* and *b* is an element of *B*, is called the Cartesian product of *A* and *B* and is denoted by *A* × *B*.

Thus a binary relation from *A* to *B* is a subset of Cartesian product *A* × *B*.

Examples: If *A* = {*1, 2, 3*} and *B* = {*4, 5*}, then {*<1, 4>, <2, 5>, <3, 5>*}, for example, is a binary relation from *A* to *B*.

However, {*<1, 1>, <1, 4>, <3, 5>*} is not a binary relation from *A* to *B* because *1* is not in *B*.

The parent–child relation is a binary relation on the set of people. *<John, John Jr.>*, for example, is an element of the parent–child relation if John is the father of John Jr.

N–ARY RELATION

Here we are going to formally define general *n*-ary relation using the concept of ordered n-tuple.

Ordered n-tuple: An ordered *n*-tuple is a set of *n* objects with an order associated with them. If *n* objects are represented by $x_1, x_2, \ldots, x_n$, then we write the ordered *n*-tuple as $<x_1, x_2, \ldots, x_n>$.

Cartesian product: Let $A_1, \ldots, A_n$ be *n* sets. Then the set of all ordered *n*-tuples $<x_1, \ldots, x_n>$, where $x_i \in A_i$ for all i, $1 \le i \le n$, is called the Cartesian product of $A_1, \ldots, A_n$, and is denoted by $A_1 \times \ldots \times A_n$.

Equality of n-tuples: Two ordered *n*-tuples $<x_1, \ldots, x_n>$ and $<y_1, \ldots, y_n>$ are equal if and only if $x_i = y_i$ for all i, $1 \le i \le n$.

For example the ordered 3–tuple *<1, 2, 3>* can be equal to only *<1, 2, 3>* and nothing else. It is not equal to the ordered *n*–tuple *<2, 3, 1>* for example.

N-ary relation: An *n*–ary relation on sets $A_1, ..., A_n$ is a set of ordered *n*–tuples $<a_1, ..., a_n>$ where a_i is an element of A_i for all i, $1 \leq i \leq n$. Thus an *n*-ary relation on sets $A_1, ..., A_n$ is a subset of Cartesian product $A_1 \times ... \times A_n$.

Example: Let A_1 be a set of names, A_2 a set of addresses, and A_3 a set of telephone numbers. Then a set of 3–tuples < name, address, telephone number > such as {< Amy Angels, 35 Mediterranean Ave, 224–1357 >, < Barbara Braves, 221 Atlantic Ave, 301–1734 >, < Charles Cubs, 312 Baltic Ave, 223–9876 >}, is a 3–ary (ternary) relation over A_1, A_2 and A_3.

Example: Let A_1 be a set of names. Then a set of 1–tuples such as {< Amy >, < Barbara >, < Charles >}, is a *1*–ary (unary) relation over A_1. A unary relation represents a property/characteristic, such as tall, rich etc., shared by the members of A_1 listed in the relation.

Special Relations

The empty set is certainly a set of ordered n–tuples. Therefore it is a relation. It is called the empty relation.

The Cartesian product $A_1 \times ... \times A_n$ of sets $A_1, ..., A_n$, is also a relation, and it is called the universal relation.

EQUALITY OF RELATIONS

Definition (equality of binary relation): Two binary relations $R_1 \subseteq A_1 \times A_2$ and $R_2 \subseteq B_1 \times B_2$ are equal if and only if $A_1 = B_1$, $A_2 = B_2$, and $R_1 = R_2$ as a set.

For example, let $R_1 = \{<1, 2>, <2, 2>\} \subseteq \{1, 2\} \times \{1, 2\}$, and $R_2 = \{<a, b>, <b, b>\} \subseteq \{a, b\} \times \{a, b\}$. Then $R_1 = R_2$ if and only if $a = 1$ and $b = 2$.

Definition (equality of n–ary relation): An *n*-ary relation $R_1 \subseteq A_1 \times ... \times A_n$ and an *m*-ary relation $R_2 \subseteq B_1 \times ... \times B_m$ are equal if and only if $m = n$, $A_i = B_i$ for each i, $1 \leq i \leq n$, and $R_1 = R_2$ as a set of ordered *n*-tuples.

RELATION

Certain relations can be defined recursively. Note that a relation is a set. Therefore a recursive definition of a relation follows the same format as that of sets. Here only examples are given.

Example: Let us define recursively the relation "less than" denoted by $R_<$ on the set of natural numbers *N*.

Note that $R_<$ = {<*a*, *b*> | *a* ∈ N ∧ *b* ∈ N ∧ *a* < *b*} = {<*0*, *1*>, <*0*, 2>,..., <1, 2>, ...}.

Basis Clause: <*0*, *1*> ∈ $R_<$ (or *0* $R_<$ 1), meaning *0* is less than *1*.

Inductive Clause: For all *x* and *y* in *N*, if <*x*, *y*> ∈ $R_<$, then <*x*, *y* + *1*> ∈ $R_<$, and <*x* + *1*, *y* + *1*> ∈ $R_<$.

Extremal Clause: Nothing is in $R_<$, unless it is obtained from the Basis and Inductive Clauses.

Informally one can see that this definition is correct as follows:

First, <*0*, *1*> is the "simplest" element in $R_<$.

Next by arranging the ordered pairs in $R_<$ as follows, one can see that the two operations in the Inductive Clause generate them all:

<0, 1>, <0, 2>, <0, 3>, ..., <0, n>, ...

<1, 2>, <1, 3>, <1, 4>, ..., <1, n>, ...

<2, 3>, <2, 4>, <2, 5>, ..., <2, n>, ...

In each row, the first (leftmost) ordered pair gives the "smallest" ordered pair with the first number. For example <*2*, *3*> is the smallest with *2* as the first component. It is obtained from the first ordered pair of the row immediately above it by using<*x* + *1*, *y* + *1*> of the Inductive Clause, apply that to <*1*, *2*> to get <*2*, *3*>, for example.

Within each row, the ordered pairs are obtained from the first one by using<*x*, *y* + *1*> of the Inductive Clause successively.

Thus all the ordered pairs of $R_<$ are generated from <*0*, *1*> by following the Inductive Clause.

Example: Let $R_{a+b=c}$ be the set of triples of natural numbers <*a*, *b*, *c*> which satisfy *a* + *b* = *c*. Then $R_{a+b=c}$ on the set of natural numbers *N* can be defined recursively as follows.

Basis Clause: <*0*, *0*, *0*> ∈ $R_{a+b=c}$.

Inductive Clause: For all *x*, *y* and *z* in *N*, if <*x*, *y*, *z*> ∈ $R_{a+b=c}$, then <*x* + *1*, *y*, *z* + *1*> and <*x*, *y* + *1*, *z* + *1*> ∈ $R_{a+b=c}$.

Extremal Clause: Nothing is in $R_{a+b=c}$ unless it is obtained from the Basis and Inductive Clauses.

PROPERTIES

DIGRAPH

A digraph is short for directed graph, and it is a diagram composed of points called vertices (nodes) and arrows called arcs going from a vertex to a vertex.

For example the figure below is a digraph with 3 vertices and 4 arcs.

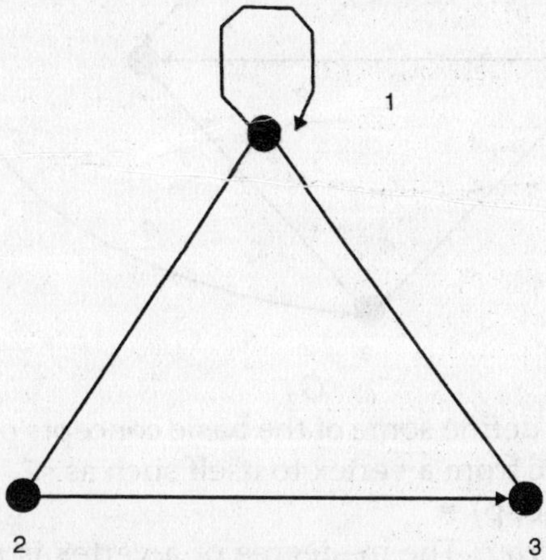

G_1

In this figure the vertices are labeled with numbers *1, 2,* and *3*.

Mathematically a digraph is defined as follows.

Digraph: A digraph is an ordered pair of sets $G = (V, A)$, where V is a set of vertices and A is a set of ordered pairs (called arcs) of vertices of V.

In the example, G_1, given above, $V = \{1, 2, 3\}$, and $A = \{\langle 1, 1\rangle, \langle 1, 2\rangle, \langle 1, 3\rangle, \langle 2, 3\rangle\}$.

Digraph representation of binary relations: A binary relation on a set can be represented by a digraph. Let R be a binary relation on a set A, that is R is a subset of $A \times A$.

Then the digraph, call it G, representing R can be constructed as follows:

1. The vertices of the digraph G are the elements of A, and
2. $\langle x, y\rangle$ is an arc of G from vertex x to vertex y if and only if $\langle x, y\rangle$ is in R.

Example: The less than relation *R* on the set of integers *A* = {*1, 2, 3, 4*} is the set {<*1, 2*>, <*1, 3*>, <*1, 4*>, <*2, 3*>, <*2, 4*>, <*3, 4*>} and it can be represented by the following digraph.

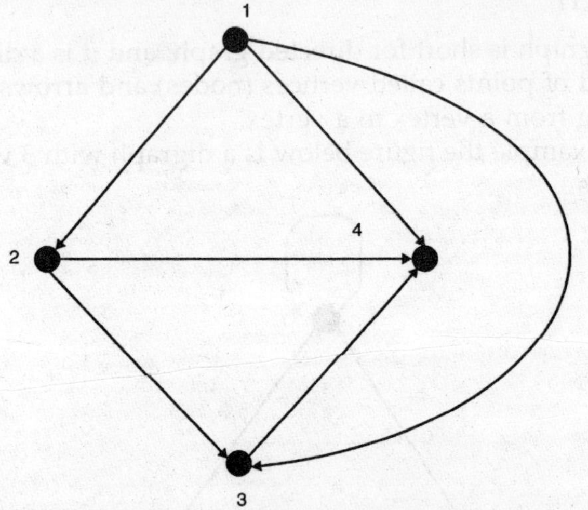

G_2

Let us now define some of the basic concepts on digraphs.

Loop: An arc from a vertex to itself such as <*1, 1*>, is called a loop (or self–loop) .

Degree of vertex: The in–degree of a vertex is the number of arcs coming to the vertex, and the out–degree is the number of arcs going out of the vertex.

For example, the in–degree of vertex 2 in the digraph G_2 shown above is *1*, and the out–degree is 2.

Path: A path from a vertex x_0 to a vertex x_n in a digraph *G* = (*V, A*) is a sequence of vertices $x_0, x_1, \ldots, x_n$ that satisfies the following:

For each $i, 0 \leq i \leq n-1$, $<x_i, x_{i+1}> \in A$, or $<x_{i+1}, x_i> \in A$, that is, between any pair of vertices there is an arc connecting them.

x_0 is the initial vertex and x_n is the terminal vertex of the path.

A path is called a directed path if $<x_i, x_{i+1}> A \in$, for every $i, 0 \leq i \leq n-1$.

If the initial and the terminal vertices of a path are the same, that is, $x_0 = x_n$, then the path is called a cycle .

If no arcs appear more than once in a path, the path is called a simple path. A path is called elementary if no vertices appear more than once in it except for the initial and terminal vertices of a cycle. In a simple cycle one vertex appears twice in the sequence: once as the initial vertex and once as the terminal vertex.

Note: There are two different definitions for "simple path". Here we follow the definition of Berge[1], Liu[2], Rosen[3] and others. A "simple path" according to another group (Cormen et al[4], Stanat and McAllister[5] and others) is a path in which no vertices appear more than once.

Connected graph: A digraph is said to be connected if there is a path between every pair of its vertices.

Example: In the digraph G_3 given below,

1, 2, 5 is a simple and elementary path but not directed,

1, 2, 2, 5 is a simple path but neither directed nor elementary.

1, 2, 4, 5 is a simple elementary directed path,

1, 2, 4, 5, 2, 4, 5 is a directed path but not simple (hence not elementary),

1, 3, 5, 2, 1 is a simple elementary cycle but not directed, and

2, 4, 5, 2 is a simple elementary directed cycle.

This digraph is connected.

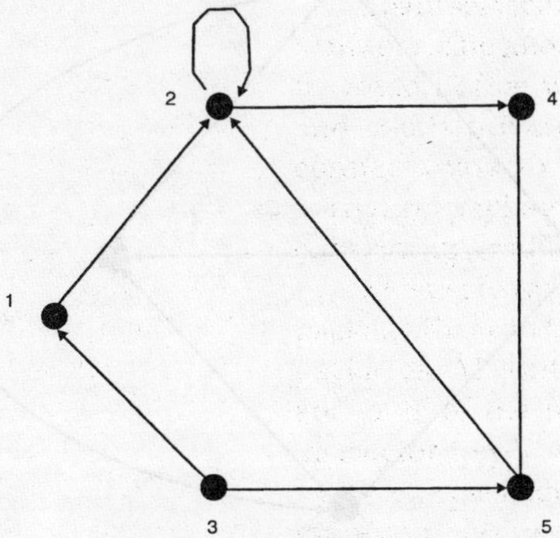

G_3

Sometimes we need to refer to part of a given digraph. A partial digraph of a digraph is a digraph consisting of arbitrary numbers of vertices and arcs of the given digraph, while a subdigraph is a digraph consisting of an arbitrary number of vertices and all the arcs between them of the given digraph. Formally they are defined as follows:

Subdigraph, partial digraph: Let $G = (V, A\}$ be a digraph. Then a digraph (V', A') is a partial digraph of G, if $V' \subseteq V$, and $A' \subseteq A \cap (V' \times V')$. It is a subdigraph of G, if $V' \subseteq V$, and $A' = A$ $(V' \times V')$.

PRESENTATION OF BINARY RELATION

A binary relation on a set can be represented by a digraph. Let R be a binary relation on a set A, that is R is a subset of $A \times A$. Then a digraph representing R can be constructed as follows:

Let the elements of A be the vertices of the digraph G, and let <*x*, *y*> be an arc of G from vertex x to vertex y if and only if <*x*, *y*> is in R.

Example: The less than relation R on th e set of integers A = {*1, 2, 3, 4*} is the set {<*1, 2*>, <*1, 3*>, <*1, 4*>, <*2, 3*>, <*2, 4*>, <*3, 4*>} and it can be represented by the following digraph.

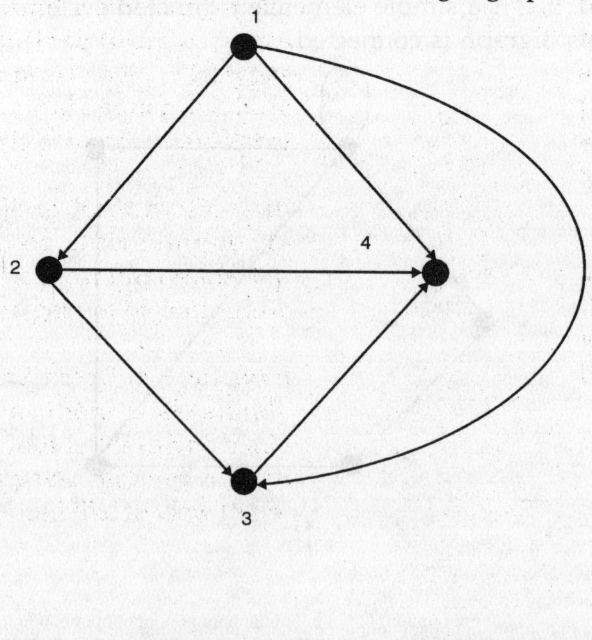

PROPERTIES OF BINARY RELATION

Certain important types of binary relation can be characterized by properties they have. Here we are going to learn some of those properties binary relations may have. The relations we are interested in here are binary relations on a set.

reflexive relation: A relation *R* on a set *A* is called reflexive if and only if < a, a > ∈*R* for every element *a* of *A*.

Example: The relation ≤ on the set of integers {*1, 2, 3*} is {<*1, 1*>, <*1, 2*>, <*1, 3*>, <*2, 2*>, <*2, 3*>, <*3, 3*>} and it is reflexive because <*1, 1*>, <*2, 2*>, <*3, 3*> are in this relation. As a matter of fact ≤ on any set of numbers is also reflexive. Similarly ≥ and = on any set of numbers are reflexive. However, < (or >) on any set of numbers is not reflexive.

Example: The relation ⊆ on the set of subsets of {1, 2} is {< Ø, Ø >, < Ø, {1} >, < Ø, {2} >, < Ø, {1, 2} >, < {1}, {1} >, < {1}, {1, 2} >, < {2}, {2} >, < {2}, {1, 2} >, < {1, 2}, {1, 2} >} and it is reflexive. In fact relation on any collection of sets is reflexive.

Irreflexive relation: A relation *R* on a set *A* is called irreflexive if and only if < *a, a* > ∉ *R* for every element *a* of *A*.

Example: The relation > (or <) on the set of integers {*1, 2, 3*} is irreflexive. In fact it is irreflexive for any set of numbers.

Example: The relation {< *1, 1* >, < *1, 2* >, < *1, 3* >, < *2, 3*>, < *3, 3* >} on the set of integers {*1, 2, 3*} is neither reflexive nor irreflexive.

Symmetric relation: A relation *R* on a set *A* is called symmetric if and only if for any *a*, and *b* in *A*, whenever <*a, b*> ∈*R*, <*b, a*> ∈*R* .

Example: The relation = on the set of integers {*1, 2, 3*} is {<*1, 1*>, <*2, 2*> <*3, 3*>} and it is symmetric. Similarly = on any set of numbers is symmetric. However, < (or >), ≤ (or ≥ on any set of numbers is not symmetric.

Example: The relation "being acquainted with" on a set of people is symmetric.

Antisymmetric relation: A relation *R* on a set *A* is called antisymmetric if and only if for any *a*, and *b* in *A*, whenever <*a, b*> ∈ *R*, and <*b, a*> ∈ *R*, *a* = *b* must hold. Equivalently, *R* is antisymmetric if and only if whenever <*a, b*> ∈ *R*, and *a* ≠ *b*,

<b, a> ∉ *R*. Thus in an antisymmetric relation no pair of elements are related to each other.

Example: The relation < (or >) on any set of numbers is antisymmetric. So is the equality relation on any set of numbers.

Transitive relation: A relation *R* on a set *A* is called transitive if and only if for any*a, b,* and *c* in *A,* whenever *<a, b>* ∈ *R,* and *<b, c>* ∈ *R, <a, c>* ∈ *R* .

Example: The relation ≤ on the set of integers {*1, 2, 3*} is transitive, because for*<1, 2>* and *<2, 3>* in ≤, *<1, 3>* is also in, for *<1, 1>* and *<1, 2>* in ≤, *<1, 2>* is also in, and similarly for the others. As a matter of fact ≤ on any set of numbers is also transitive. Similarly and = ≥ on any set of numbers are transitive.

The following figures show the digraph of relations with different properties.

1. Is reflexive, antisymmetric, symmetric and transitive, but not irreflexive.
2. Is neither reflexive nor irreflexive, and it is antisymmetric, symmetric and transitive.
3. Is irreflexive but has none of the other four properties.
4. Is irreflexive, and symmetric, but none of the other three.
5. Is irreflexive, antisymmetric and transitive but neither reflexive nor symmetric.

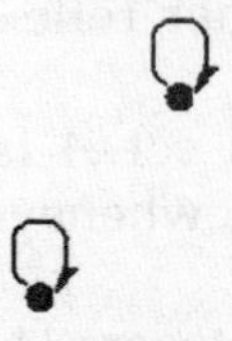

(a)

(b)

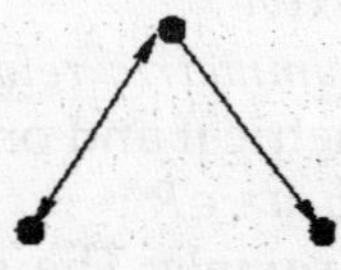

(c)

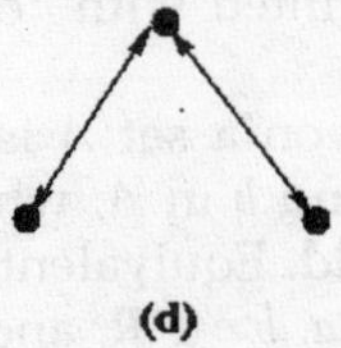

(d)

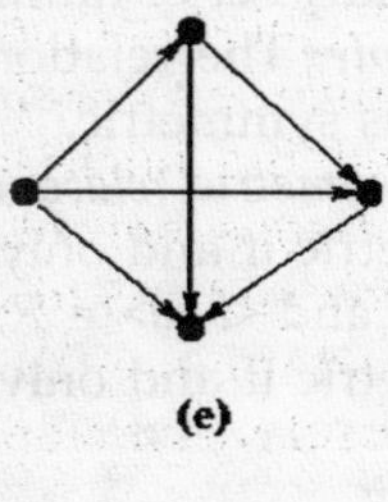

(e)

OPERATIONS ON RELATIONS

SET OPERATIONS

A relation is a set. It is a set of ordered pairs if it is a binary relation, and it is a set of ordered *n*–tuples if it is an *n*–ary relation. Thus all the set operations apply to relations such as $\cup$, $\cap$, and complementing.

For example, the union of the "less than" and "equality" relations on the set of integers is the "less than or equal to" relation on the set of integers. The intersection of the "less than" and "less than or equal to" relations on the set of integers is the "less than" relation on the same set. The complement of the "less than" relation on the set of integers is the "greater than or equal to" relation on the same set.

RELATIONS

If the elements of a set *A* are related to those of a set *B*, and those of *B* are in turn related to the elements of a set *C*, then one can expect a relation between *A* and *C*. For example, if Tom is my father(parent–child relation) and Sarah is a sister of Tom (sister relation), then Sarah is my aunt (aunt–nephew/niece relation). Composite relations give that kind of relations.

Composite relation: Let R_1 be a binary relation from a set *A* to a set *B*, R_2 a binary relation from *B* to a set *C*. Then the composite relation from *A* to *C* denoted by R_1R_2(also denoted by R_1 Ï R_2 is defined as

$$R_1R_2 = \{<a, c> \mid a \in A \wedge c\ C \wedge \exists\, b\ [b \in B < \wedge\ a, b> \in R_1 \wedge <b, c> \in R_2]\}.$$

In English, this means that an element *a* in *A* is related to an element *c* in *C* if there is an element *b* in *B* such that *a* is related to *b* by R_1 and *b* is related to *c* by R_2 . Thus R_1R_2 is a reiation from *A* to *C* via *B* in a sense. If R_1 is a parent–child relation and R_2 is a sister relation, then R_1R_2 is an aunt–nephew/niece relation.

Example: Let $A = \{a_1, a_2\}$, $B = \{b_1, b_2, b_3\}$, and $C = \{c_1, c_2\}$. Also let $R_1 = \{<a_1, b_1>, <a_1, b_2>, <a_2, b_3>\}$, and $R_2 = \{<b_1, c_1>, <b_2, c_1>, <b_2, c_2>, <b_3, c_1>\}$. Then $R_1R_2 = \{<a_1, c_1>, <a_1, c_2>, <a_2, c_1>\}$.

This is illustrated in the following figure. The dashed lines in the figure of R_1R_2 indicate the ordered pairs in R_1R_2, and dotted lines show ordered pairs that produce the dashed lines. (The lines in the left figure are all supposed to be solid lines.)

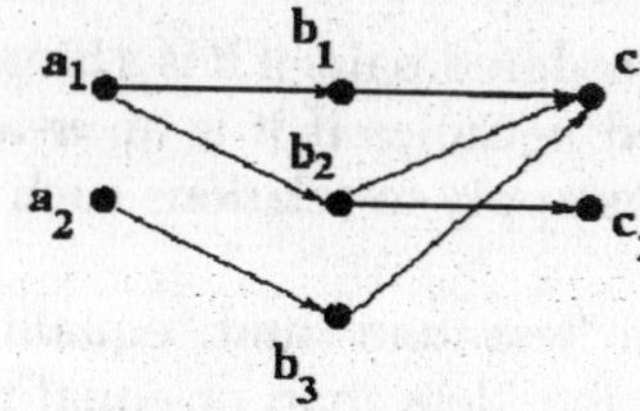

R_1 R_2

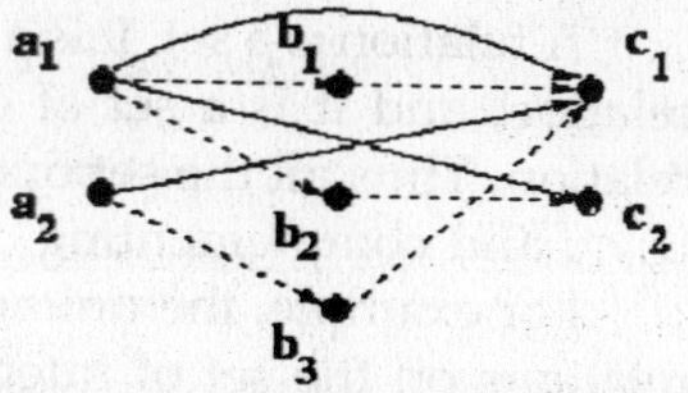

R_1R_2

Example: If R is the parent–child relation on a set of people A, then RR, also denoted by R^2, is the grandparent–grandchild relation on A.

More examples: The digraphs of R^2 for several simple relations R are shown below:

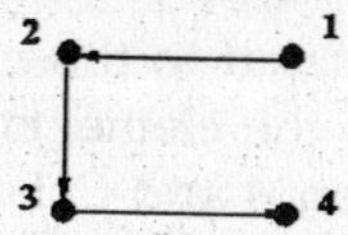

R_1

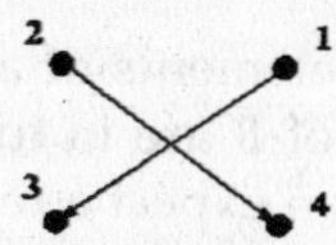

R_1^2

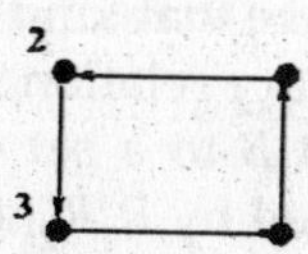

R_2

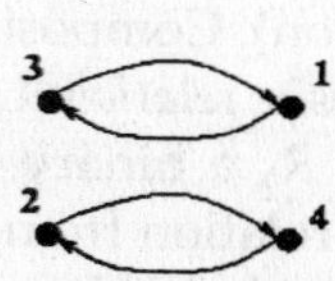

R_2^2

R_3

R_3^2

PROPERTIES OF COMPOSITE RELATIONS

Composite relations defined above have the following properties. Let R_1 be a relation from A to B, and R_2 and R_3 be relations from B to C. Then

1. $R_1(R_2R_3) = (R_1R_2)R_3$

2. $R_1(R_2 \cup R_3) = R_1R_2 \cup R_1R_3$
3. $R_1(R_2 \cap R_3) \subseteq R_1R_2 \cap R_1R_3$

Proofs for these properties are omitted.

POWERS OF RELATION

Let R be a binary relation on A. Then R^n for all positive integers n is defined recursively as follows:

Definition(power of relation):

Basis Clause: $R^0 = E$, where E is the equality relation on A.

Inductive Clause: For an arbitrary natural number n, $R^{n+1} = R^nR$.

Note that there is no need for extremal clause here.

Thus for example $R^1 = R$, $R^2 = RR$, and $R^3 = R^2R = (RR)R = R(RR) = RRR$.

The powers of binary relation R on a set A defined above have the following properties.

1. $R^{m+n} = R^mR^n$,
2. $(R^m)^n = R^{mn}$.

SPECIAL RELATIONS

BINARY RELATION

In our everyday life we often talk about parent-child relationship. This is a binary relation on the set of people in the world, dead or alive. Also we are often interested in ancestor-descendant relations. Although the latter relation can be obtained from the former, hence it is redundant in that sense, we do use ancestor-descendant relations which give us necessary information more directly. This ancestor-descendant relation relates two people if there is a sequence of parent-child relations from one to the other.

It includes the parent-child relation as a subset. The ancestor-descendant relation is an example of the closure of a relation, in particular the transitive closure of the parent-child relation. In general, the closure of a relation is the smallest extension of the relation that has a certain specific property such as the reflexivity, symmetry or transitivity. Formally they are defined as follows:

Reflexive closure: A relation R' is the reflexive closure of a relation R if and only if

1. R' is reflexive,
2. $R R'$, and
3. For any relation R'', if $R \subseteq R''$ and R'' is reflexive, then $R' \subseteq R''$, that is, R' is the smallest relation that satisfies (1) and (2).

Example: Let R be the less-than relation on the set of integers I, that is $R = \{<a, b> \mid a \in I \wedge b \in I \wedge a < b\}$.

Then the reflexive closure $r(R)$ of R is the union of R and the equality relation on I, that is $r(R) = \{<a, b> \mid a \in I \wedge b \in I\ a \in b\}$

The digraph of the reflexive closure of a relation is obtained from the digraph of the relation by adding a self-loop at each vertex if one is already not there.

Symmetric and transitive closures can be defined similarly.

Symmetric closure: A relation R' is the symmetric closure of a relation R if andonly if

1. R' is symmetric,
2. $R R'$, and
3. For any relation R'', if $R \subseteq R''$, and R'' is symmetric, then $R' \subseteq R''$, that is, R' is the smallest relation that satisfies (1) and (2).

Example: Let R be the less–than relation on the set of integers I. Then the symmetric closure of R, denoted by $s(R)$ is $s(R) = \{<a, b> \mid a\ I\ b\ I\ [a < b\ a > b]\}$ that is $\{<a, b> \mid a\ I\ b\ I\ a\ b\}$

The digraph of the symmetric closure of a relation is obtained from the digraph of the relation by adding for each arc the arc in the reverse direction if one is already not there.

Transitive closure: A relation R' is the transitive closure of a relation R if and only if

1. R' is transitive,
2. $R \subseteq R'$, and
3. For any relation R'', if $R \subseteq R''$ and R'' is transitive, then $R' \subseteq R''$, that is, R' is the smallest relation that satisfies (1) and (2).

PROPERTIES OF CLOSURE

The closures have the following properties. They are stated here as theoremswithout proof.

Theorem: Let *E* denote the equality relation, and R^c the inverse relation of binary relation *R*, all on a set *A*, where $R^c = \{< a, b > \mid < b, a \in R\}$. Then

1. $r(R) = R \cup E$
2. $s(R) = R \cup R^c$
3. $t(R) = \cup_{i=1}^{\infty} R^i = \cup_{i=1}^{n} R^i$, if $|A| = n$.
4. *R* is reflexive if and only if $r(R) = R$.
5. *R* is symmetric if and only if $s(R) = R$.
6. *R* is transitive if and only if $t(R) = R$.

EQUIVALENCE RELATION

On the face of most clocks, hours are represented by integers between 1 and 12. However, since a day has 24 hours after 12 hours, a clock goes back to hour 1, and starts all over again from there. Thus each pair of hours such as 1 and 13, 2 and 14, etc. share one number 1, 2, ...etc., respectively.

The same applies when we are interested in more than 24 hours. 25th hour is 1, so are 37th, 49th etc. What we are doing here essentially is that we consider the numbers in each group such as 1, 13, 25, ..., equivalent in the sense that they all are represented by one number. Being representable by one number such as we see on clocks is a binary relation on the set of natural numbers and it is an example of equivalence relation we are going to study here.

The concept of equivalence relation is characterized by three properties as follows:

Definition(equivalence relation): A binary relation *R* on a set *A* is an equivalence relation if and only if

- *R* is reflexive
- *R* is symmetric, and
- *R* is transitive.

Example: The equality relation (=) on a set of numbers such as {*1, 2, 3*} is an equivalence relation.

Example: The congruent modulo *m* relation on the set of integers i.e. {<*a*, *b*>| *a b* (mod *m*)}, where *m* is a positive integer greater than *1*, is an equivalence relation.

Note that the equivalence relation on hours on a clock is the congruent mod *12*, and that when *m* = 2, i.e. the congruent mod *2*, all even numbers are equivalent and all odd numbers are equivalent. Thus the set of integers are divided into two subsets: evens and odds.

Example: Taking this discrete structures course together this semester is another equivalence relation.

Equivalence relations can also be represented by a digraph since they are a binary relation on a set. For example the digraph of the equivalence relation congruent mod *3* on {*0, 1, 2, 3, 4, 5, 6*} is as shown below. It consists of three connected components.

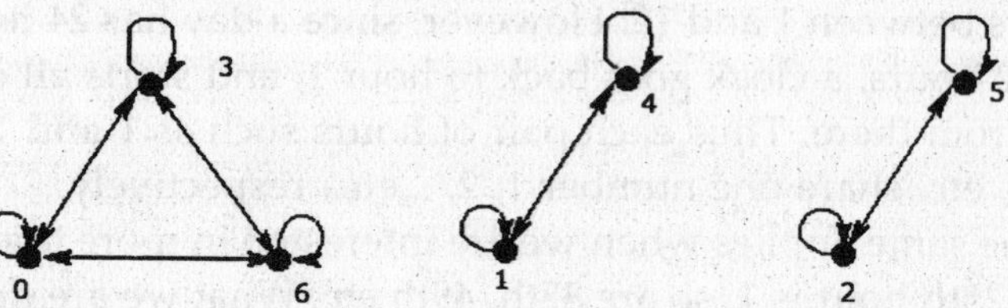

The set of even numbers and that of odd numbers in the equivalence relation of congruent mod 2, and the set of integers equivalent to a number between 1 and 12 in the equivalence relation on hours in the clock example are called an equivalence class. Formally it is defined as follows:

Equivalence class: For an equivalence relation *R* on a set *A*, the set of the elements of *A* that are related to an element, say *a*, of *A* is called the equivalence class of element *a* and it is denoted by [*a*].

Example: For the equivalence relation of hours on a clock, equivalence classes are

[*1*] = {*1, 13, 25, ...*} = {*1+ 12n*: *n* ∈*N*},

[2] = {*2, 14, 26, ...*} = {*2+ 12n*: *n* ∈*N*},

........,

where *N* is the set of natural numbers. There are altogether twelve of them.

For an equivalence relation R on a set A, every element of A is in an equivalence class. For if an element, say b, does not belong to the equivalence class of any other element in A, then the set consisting of the element b itself is an equivalence class. Thus the set A is in a sense *covered* by the equivalence classes.

Another property of equivalence class is that equivalence classes of two elements of a set A are either disjoint or identical, that is either $[a] = [b]$ or $[a] \cap [b] = \emptyset$ for arbitrary elements a and b of A.

Thus the set A is partitioned into equivalence classes by an equivalence relation on A. This is formally stated as a theorem below after the definition of partition.

Partition: Let A be a set and let $A_1, A_2, \ldots, A_n$ be subsets of A. Then $\{A_1, A_2, \ldots, A_n\}$ is a partition of A, if and only if

1. $\cup_{i=1}^{n} A_i = A$, and
2. $A_i \cap A_j = \emptyset$, if $A_i \neq A_j$, $1 \leq i, j \leq n$.

Example: Let $A = \{1, 2, 3, 4, 5\}$, $A_1 = \{1, 5\}$, $A_2 = \{3\}$, and $A_3 = \{2, 4\}$. Then $\{A_1, A_2, A_3\}$ is a partition of A. However, $B_1 = \{1, 2, 5\}$, $B_2 = \{2, 3\}$, and $B_3 = \{4\}$ do not form a partition for A because $B_1 \cap B_2 \neq \emptyset$, though $B_1 \neq B_2$.

Theorem: The set of equivalence classes of an equivalence relation on a set A is a partition of A.

Conversely, a partition of a set A determines an equivalence relation on A.

Theorem: Let $\{A_1, \ldots, A_n\}$ be a partition of a set A. Define a binary relation R on A as follows: $<a, b> \in R$ if and only if $a \in A_i$ and $b \in A_i$ for some i, $1 \leq i \leq n$. Then R is an equivalence relation.

Theorem : Let R_1 and R_2 be equivalence relations. Then $R_1 \cap R_2$ is an equivalence relation, but $R_1 \cup R_2$ is not necessarily an equivalence relation.

ORDER OF RELATION

The properties common to orders we see in our daily lives have been extracted and are used to characterize the concepts of order. Here we are going to learn three types of order: partial order, total order, and quasi order.

Partial order: A binary relation *R* on a set *A* is a partial order if and only if it is :

1. Reflexive,
2. Antisymmetric, and
3. Transitive.

The ordered pair <*A, R*> is called a poset (partially ordered set) when *R* is a partial order.

Example: The less-than-or-equal-to relation on the set of integers *I* is a partial order, and the set *I* with this relation is a poset.

Example: The subset relation on the power set of a set, say {1, 2}, is also a partial order, and the set {1, 2} with the subset relation is a poset.

Total order: A binary relation *R* on a set *A* is a total order if and only if it is

1. A partial order, and
2. For any pair of elements *a* and *b* of *A*, $< a, b > \in R$ or $< b, a > \in R$.

That is, every element is related with every element one way or the other.

A total order is also called a *linear order*.

Example: The less-than-or-equal-to relation on the set of integers *I* is a total order.

The strictly-less-than and proper-subset relations are not partial order because they are not reflexive. They are examples of some relation called quasi order.

Quasi order: A binary relation *R* on a set *A* is a quasi order if and only if it is

1. Irreflexive, and
2. Transitive.

A quasi order is necessarily antisymmetric as one can easily verify.

Caution Like many other definitions there is another fairly widely used definition of quasi order in the literature. According to that definition a quasi order is a relation that is *reflexive* and transitive. That is something quite different from "quasi order" we have defined here.

Example: The less-than relation on the set of integers I is a quasi order.

Example: The proper subset relation on the power set of a set, say {1, 2}, is also a quasi order.

The concept of *least/greatest number* in a set of integers can be generalized for a general poset. We start with the concepts of minimal/maximal elements.

minimal/maximal element: Let $< A, \preceq >$ be a poset, where $\preceq$ represents an arbitrary partial order. Then an element $b \in A$ is a minimal element of A if there is no element $a \in A$ that satisfies $a \preceq b$. Similarly an element $b \in A$ is a maximal element of A if there is no element $a \in A$ that satisfies $b \preceq a$.

Example: The set of {{1}, {2}, {1, 2}} with $\subseteq$ has two minimal elements {1} and {2}. Note that {1}, and {2} are not related to each other in $\subseteq$. Hence we can not say which is "smaller than" which, that is, they are not comparable.

least/greatest element: Let $< A, \preceq >$ be a poset. Then an element $b \in A$ is the least element of A if for every element $a \in A$, $b \preceq a$.

Note that the least element of a poset is unique if one exists because of the antisymmetry of $\preceq$.

Example: The poset of the set of natural numbers with the less–than–or–equal–to relation has the least element 0.

Example: The poset of the powerset of {1, 2} with $\in$ has the least element Ø.

Well order: A total order R on a set A is a well order if every non–empty subset of A has the least element.

Example: The poset of the set of natural numbers with the less–than–or–equal–to relation is a well order, because every set of natural numbers has the least element.

The poset of the set of positive real numberes with the less–than–or–equal–to relation is not a well order, because the set itself does not have any least element (0 is not in the set).

A digraph of a binary relation on a set can be simplified if the relation is a partial order. Hasse diagrams defined as follows are such graphs.

Hasse diagram: A Hasse diagram is a graph for a poset

which does not have loops and arcs implied by the transitivity. Further, it is drawn so that all arcs point upward eliminating arrowheads.

To obtain the Hassse diagram of a poset, first remove the loops, then remove arcs < a, b > if and only if there is an element c that < a, c > and < c, b > exist in the given relation.

Example: For the relation {< *a, a* >, < *a, b* >, < *a, c* >, < *b, b* >, < *b, c* >, < *c, c* >} on set {*a, b,c*}, the Hasse diagram has the arcs {< *a, b* >, < *b, c* >} as shown below.

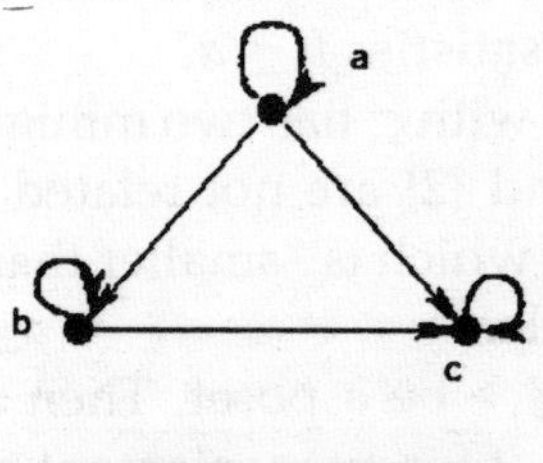

Digraph for Portial Order

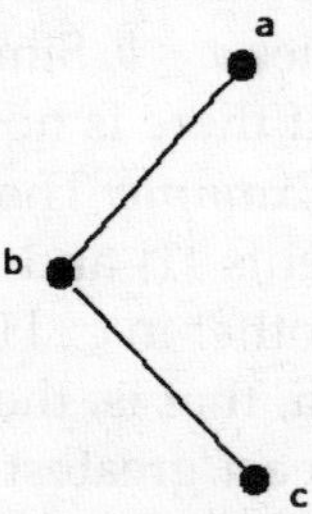

Hasse Digram

TOPOLOGICAL SORTING

The elements in a finite poset can be ordered linearly in a number of ways while preserving the partial order. For example {Ø, {1}, {2}, {1, 2}} wit h the partial order ⊆, can be ordered linearly as Ø, {1}, {2}, {1, 2}, or Ø, {2}, {1}, {1, 2}. In these orders a set appears before (to the left of) another set if it is a subset of the other. In real life, tasks for manufacturing goods in general can be partially ordered based on the prerequisite relation, that is certain tasks must be completed before certain other tasks can be started. For example the arms of a chair must be carved before the chair is assembled. Scheduling those tasks is essentially the same as arranging them with a linear order (ignoring here some possible concurrent processing for simplicity's sake).

The topological sorting is a procedure to find from a partial order on a finite set a linear order that does not violate the partial order. It is based on the fact that a finite poset has at least one minimal element. The basic idea of the topological sorting is to first remove a minimal element from the given poset, and then repeat that for the resulting set until no more elements are left. The order of removal of the minimal elements gives a linear order. The following algorithm formally describes the topological sorting.

Algorithm Topological Sor

Input: A finite poset <*A*, *R*>.

Output: A sequence of the elements of *A* preserving the order *R*.

integer *i*;
i : = 1;
while (*A*) {
pick a minimal element *b* from *A*;
A : = *A* – {*b*};
i : = *i* + 1;
output *b*
}

Example: Let *A* = {{1}, {2}, {3}, {1, 2}, {1, 3}, {2, 3}, {1, 2, 3}} with the partial order ⊆.

This given *A* has three minimal elements {1}, {2}, and {3}.

Select {2} and remove it from *A*. Let *A* denote the resultant set i.e. *A* := *A* – {2}. The new *A* has two minimal elements {1}, and {3}.

Select {1} and remove it from *A*. Denote by *A* the resultant set, that is *A* = {{3}, {1, 2}, {1, 3}, {2, 3}, {1, 2, 3}}.

This new *A* has two minimal elements {3} and {1, 2}.

Select {1, 2} and remove it from *A*.

Proceeding in like manner, we can obtain the following linear order:

{{2}, {1}, {1, 2}, {3}, {1, 3}, {2, 3}, {1, 2, 3}}.

FUNCTION

A function is something that associates each element of a set with an element of another set (which may or may not be the same as the first set). The concept of function appears quite often even in nontechnical contexts. For example, a social security number uniquely identifies the person, the income tax rate varies depending on the income, the final letter grade for a course is often determined by test and exam scores, homeworks and projects, and so on.

In all these cases to each member of a set (social security number, income, tuple of test and exam scores, homeworks and projects) some member of another set (person, tax rate, letter grade, respectively) is assigned.

As you might have noticed, a function is quite like a relation. In fact, formally, we define a function as a special type of binary relation.

Definition (function): A function, denote it by *f*, from a set *A* to a set *B* is a relation from *A* to *B* that satisfies:

1. For each element *a* in *A*, there is an element *b* in *B* such that <*a*, *b*> is in the relation, and
2. If <*a*, *b*> and <*a*, *c*> are in the relation, then $b = c$.

The set *A* in the above definition is called the domain of the function and *B* its codomain.

Thus, *f* is a function if it *covers* the domain (maps every element of the domain) and it is *single valued*.

The relation given by *f* between *a* and *b* represented by the ordered pair <*a*, *b*> is denoted as $f(a) = b$, and *b* is called the image of *a* under *f* .

The set of images of the elements of a set *S* under a function *f* is called the image of the set *S* under *f*, and is denoted by *f(S)*, that is,

$f(S) = \{f(a) \mid a\ S\}$, where *S* is a subset of the domain *A* of *f*.

The image of the domain under *f* is called the range of *f* .

Example: Let *f* be the function from the set of natural numbers *N* to *N* that maps each natural number *x* to x^2 . Then the domain and codomain of this *f* are *N*, the image of, say *3*, under this function is *9*, and its range is the set of squares, i.e. $\{0, 1, 4, 9, 16, ...\}$.

Sum and product: Let *f* and *g* be functions from a set *A* to the set of real numbers *R*.

Then the sum and the product of *f* and *g* are defined as follows:

For all *x*, $(f + g)(x) = f(x) + g(x)$, and

for all x, $(f^*g)(x) = f(x)^*g(x)$,

where $f(x)^*g(x)$ is the product of two real numbers $f(x)$ and $g(x)$.

Example: Let $f(x) = 3x + 1$ and $g(x) = x^2$. Then $(f + g)(x) = x^2 + 3x + 1$, and$(f^*g)(x) = 3x^3 + x^2$

One-to-one: A function f is said to be one–to–one (injective), if and only if whenever $f(x) = f(y)$, $x = y$.

Example: The function $f(x) = x^2$ from the set of natural numbers N to N is a one–to–one function. Note that $f(x) = x^2$ is not one–to–one if it is from the set of integers(negative as well as non–negative) to N, because for example $f(1) = f(-1) = 1$.

Onto: A function f from a set A to a set B is said to be onto(surjective), if and only if for every element y of B, there is an element x in A such that $f(x) = y$, that is, f is onto if and only if $f(A) = B$.

Example: The function $f(x) = 2x$ from the set of natural numbers N to the set ofnon–negative even numbers E is an onto function. However, $f(x) = 2x$ from the set of natural numbers N to N is not onto, because, for example, nothing in N can be mapped to 3 by this function.

Bijection: A function is called a bijection, if it is onto and one–to–one.

Example: The function $f(x) = 2x$ from the set of natural numbers N to the set ofnon–negative even numbers E is one–to–one and onto. Thus it is a bijection.

Every bijection has a function called the inverse function.

These concepts are illustrated in the figure below. In each figure below, the points on the left are in the domain and the ones on the right are in the codomain, and arrows show $< x, f(x) >$ relation.

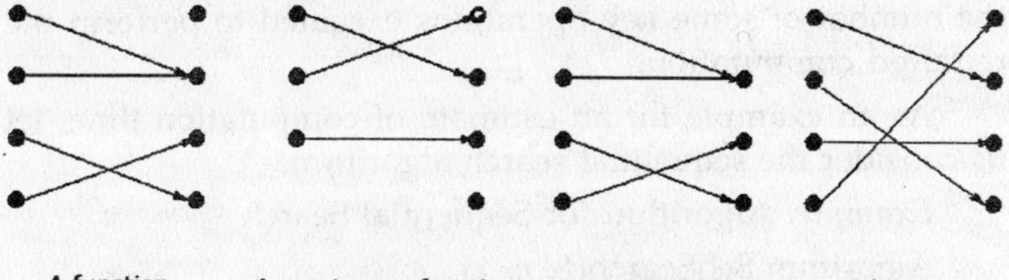

A function | A one-to-one function (Not onto) | An onto function Not one-to-one) | A bijection

Inverse: Let f be a bijection from a set A to a set B. Then the function g is called the inverse function of f, and it is denoted by f^{-1}, if for every element y of B, $g(y) = x$, where $f(x) = y$. Note that such an x is unique for each y because f is a bijection.

For example, the rightmost function in the above figure is a bijection and its inverse is obtained by reversing the direction of each arrow.

Example: The inverse function of $f(x) = 2x$ from the set of natural numbers N to the set of non–negative even numbers E is $f^{-1}(x) = 1/2\ x$ from E to N . It is also a bijection.

A function is a relation. Therefore one can also talk about composition of functions.

Composite function: Let g be a function from a set A to a set B, and let f be a function from B to a set C . Then the composition of functions f and g, denoted by fg, is the function from A to C that satisfies

$fg(x) = f(g(x))$ for all x in A .

Example: Let $f(x) = x^2$, and $g(x) = x + 1$. Then $f(g(x)) = (x + 1)^2$.

GROWTH OF FUNCTIONS

One of the important criteria in evaluating algorithms is the time it takes to complete a job. To have a meaningful comparison of algorithms, the estimate of computation time must be independent of the programming language, compiler, and computer used; must reflect on the size of the problem being solved; and must not depend on specific instances of the problem being solved. The quantities often used for the estimate are the worst case execution time, and average execution time of an algorithm, and they are represented by the number of some key operations executed to perform the required computation.

As an example for an estimate of computation time, let us consider the sequential search algorithm.

Example: Algorithm for Sequential Search

Algorithm SeqSearch(L, n, x)

L is an array with *n* entries indexed *1, .., n,* and *x* is the key to be searched for in *L.*

Output: if *x* is in *L,* then output its index, else output *0.*

Index := *1;*

while (*index* $\leq$ *n* and *L*[*index*] $\neq$ *x*)

Index := *index* + *1* ;

if (*index* > *n*), then *index* := *0*

return *index* .

The worst case time of this algorithm, for example, can be estimated as follows: First the key operation is comparison of keys comparing *L*[*index*] with *x* . Most search algorithms (if not all) need "comparison of keys". The largest number of execution of this comparison is *n,* which occurs when *x* is not in *L* or when *x* is at *L*[*n*], and the while loop is executed *n* times. This quantity *n* thus obtained is used as an estimate of the worst case time of this sequential search algorithm.

Note that in the while loop two comparisons and one addition are performed. Thus one could use *3n* as an estimate just as well. Also note that the very first line and the last two lines are not counted in. The reasons for those are firstly that differences in implementation details such as languages, commands, compilers and machines make differences in constant factors meaningless, and secondly that for large values of *n,* the highest degree term in *n* dominates the estimate. Since we are mostly interested in the behaviour of algorithms for large values of *n,* lower terms can be ignored compared with the highest term. The concept that is used to address these issues is something called big–oh, and that is what we are going to study here.

BIG – OH

The following example gives the idea of one function growing more rapidly than another. We will use this example to introduce the concept the big–Oh.

Example: $f(n) = 100\ n^2$, $g(n) = n^4$, the following table and figure show that *g(n)* grows faster than *f(n)* when *n* > 10. We say *f* is big–Oh of *g*.

n	$f(n)$	$g(n)$
10	10,000	10,000
50	250,000	6,250,000
100	1,000,000	100,000,000
150	2,250,000	506,250,000

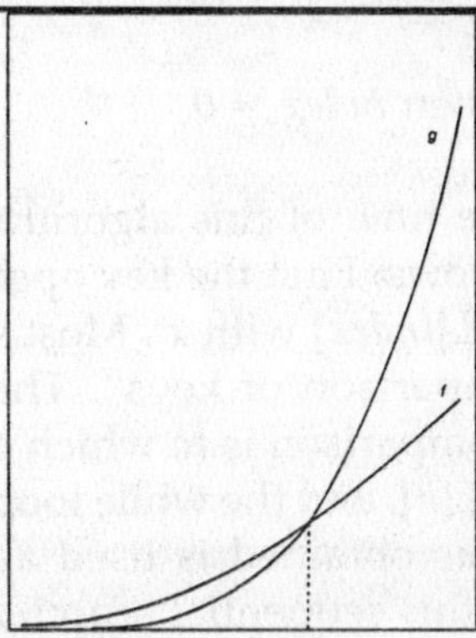

Big-oh: Let f and g be functions from the set of integers (or the set of real numbers) to the set of real numbers. Then $f(x)$ is said to be $O(g(x))$, which is read as $f(x)$ is big-oh of $g(x)$, if and only if there are constants C and n_0 such that

$$| f(x) | \; C \; | g(x) |$$

whenever $x > n_0$.

Note that big–oh is a binary relation on a set of functions (What kinds of properties does it have ? reflexive ? symmetric ? transitive ?).

For example, $5x + 10$ is big–oh of x^2, because $5x + 10 < 5x^2 + 10x^2 = 15x^2$ for $x > 1$.

Hence for $C = 15$ and $n_0 = 1$, $| 5x + 10 | \leq C \, | x^2 |$.

Similarly it can be seen that $3x^2 + 2x + 4 < 9x^2$ for $x > 1$. Hence $3x^2 + 2x + 4$ is $O(x^2)$.

In general, we have the following theorem:

Theorem: $a_n x^n + ... + a_1 x + a_0$ is $O(x^n)$ for any real numbers $a_n, ..., a_0$ and any nonnegative number n .

Note: Let $f(x) = 3x^2 + 2x + 4$, $g(x) = x^2$, from the above illustration, we have that $f(x)$ is $O(g(x))$. Also, since $x^2 < 3x^2 + 2x + 4$, we can also get $g(x)$ is $O(f(x))$. In this case, we say these two functions are of the *same order*. Please refer to *big–theta* for more information.

GROWTH OF COMBINATIONS OF FUNCTIONS

Big–oh has some useful properties. Some of them are listed as theorems here.

Let use start with the definition of max function.

Definition(max function): Let $f_1(x)$ and $f_2(x)$ be functions from a set A to a set of real numbers B. Then $\max(f_1(x), f_2(x))$ is the function from A to B that takes as its value at each point x the larger of $f_1(x)$ and $f_2(x)$.

Theorem: If $f_1(x)$ is $O(g_1(x))$, and $f_2(x)$ is $O(g_2(x))$, then $(f_1 + f_2)(x)$ is $O(\max(g_1(x), g_2(x)))$.

From this theorem it follows that if $f_1(x)$ and $f_2(x)$ are $O(g(x))$, then $(f_1 + f_2)(x)$ is $O(g(x))$, and

$(f_1 + f_2)(x)$ is $O(\max(f_1(x), f_2(x)))$.

Theorem: If $f_1(x)$ is $O(g_1(x))$, and $f_2(x)$ is $O(g_2(x))$, then $(f_1 * f_2)(x)$ is $O(g_1(x) * g_2(x))$.

Chapter 13

Introduction to Matrix

BACKGROUND

In mathematics, a matrix (plural matrices) is a rectangular table of *elements* (or *entries*), which may be numbers or, more generally, any abstract quantities that can be added and multiplied. Matrices are used to describe linear equations, keep track of the coefficients of linear transformations and to record data that depend on multiple parameters. Matrices are described by the field of matrix theory. They can be added, multiplied, and decomposed in various ways, which also makes them a key concept in the field of linear algebra.

In this article, the entries of a matrix are real or complex numbers unless otherwise noted.

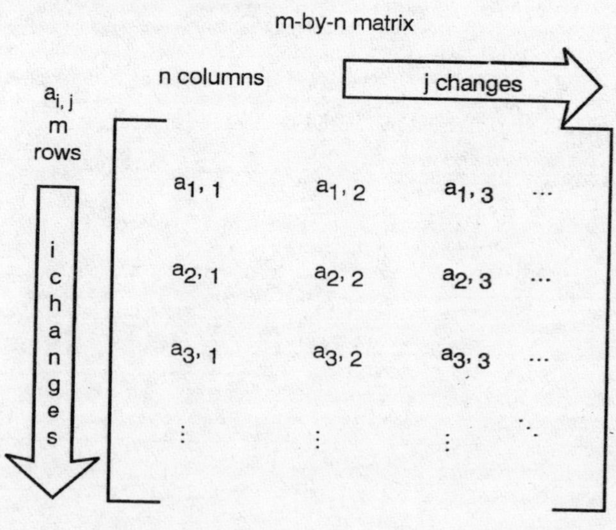

TERMS AND NOTATION

The horizontal lines in a matrix are called rows and the vertical lines are called columns. A matrix with m rows and n columns is called an m-by-n matrix (written $m \times n$) and m and n are called its dimensions. The dimensions of a matrix are always given with the number of rows first, then the number of columns. It is commonly said that an m-by-n matrix has an order of $m \times n$ ("order" meaning size). Two matrices of the same order whose corresponding entries are equivalent are considered equal.

The entry that lies in the i-th row and the j-th column of a matrix is typically referred to as the i,j, or (i,j), or (i,j)-th, or $(i,j)^{th}$ entry of the matrix. Again, the row is always noted first, then the column.

Almost always upper-case letters denote matrices, while the corresponding lower-case letters, with two subscript indices, represent the entries. For example, the $(i,j)^{th}$ entry of a matrix A is most commonly written as $a_{i,j}$. Alternative notations for that entry are $A[i,j]$ or $A_{i,j}$. In addition to using upper-case letters to symbolize matrices, many authors use a special typographical style, commonly boldface upright (non-italic), to further distinguish matrices from other variables. Following this convention, A is a matrix, distinguished from *A*, a scalar. An alternate convention is to annotate matrices with their dimensions in small type underneath the symbol, for example, $A_{m \times n}$ for an m-by-n matrix.

We often write $A := (a_{i,j})_{i=1,m;\ j=1,\dots,n}$ or $A := (a_{i,j})_{m \times n}$ to define an $m \times n$ matrix A. In this case, the entries $a_{i,j}$ are defined separately for all integers $1 \le i \le m$ and $1 \le j \le n$. In some programming languages, the numbering of rows and columns starts at zero. Texts which use any such language extensively frequently follow that convention, so we have $0 \le i \le m-1$ and $0 \le j \le n-1$.

A matrix where one of the dimensions equals one is often called a *vector*, and interpreted as an element of real coordinate space. An $m \times 1$ matrix (one column and m rows) is called a column vector and a $1 \times n$ matrix (one row and n columns) is called a row vector.

DEFINITION

An $m \times n$ ($m, n \in \mathbb{N}$) matrix A is a function $A : \{1, 2, ..., m\} \times \{1, 2, ..., n\} \rightarrow S$, where S is any non-empty set. ($\{1, 2, ..., m\} \times \{1, 2, ..., n\}$ is the Cartesian product of sets $\{1, 2, ..., m\}$ and ($\{1, 2, ..., n\}$.)

We say that matrix A is a matrix over the set . Important thing to note is that, if we want to have matrix algebra, the set must be a ring and matrix must be a square matrix (see Square matrices and related definitions below for further explanation). Since the set of all square matrices over a ring is also a ring, matrix algebra is usually called matrix ring.

Since this article mainly considers matrices over real numbers, matrices shown here are actually functions A : ($\{1, 2, ..., m\} \times \{1, 2, ..., n\} \rightarrow \mathbb{R}$.

Example: The matrix

$$A = \begin{bmatrix} 9 & 8 & 6 \\ 1 & 2 & 7 \\ 4 & 9 & 2 \\ 6 & 0 & 5 \end{bmatrix} \text{ or } A = \begin{pmatrix} 9 & 8 & 6 \\ 1 & 2 & 7 \\ 4 & 9 & 2 \\ 6 & 0 & 5 \end{pmatrix}$$

is a 4×3 matrix. The element $a_{2,3}$ or A [2, 3] is 7. In terms of the mathematical definition given above, this matrix is a function A: $\{1, 2, 3, 4\} \times \{1, 2, 3\} \rightarrow \mathbb{R}$ and, for example, $A((2, 3)) = 7$ and $A((3, 1)) = 4$.

The matrix

$$R = \begin{bmatrix} 1 & 2 & 3 & 4 & 5 & 6 & 7 & 8 & 9 \end{bmatrix}$$

is a 1×9 matrix, or 9-element row vector.

BASIC OPERATIONS

SUM

Two or more matrices of identical dimensions m and n

can be added. Given *m*-by-*n* matrices A and B, their sum A+B is the *m*-by-*n* matrix computed by adding corresponding elements:

$$A + B = (a_{i,j})_{1\le i\le m;\,1\le j\le n} + (b_{i,j})_{1\le i\le m;\,1\le j\le n}$$
$$= (a_{i,j} + b_{i,j})_{1\le i\le m;\,1\le j\le n}\,.$$

For example:

$$\begin{bmatrix}1 & 3 & 1\\ 1 & 0 & 0\\ 1 & 2 & 2\end{bmatrix} + \begin{bmatrix}0 & 0 & 5\\ 7 & 5 & 0\\ 2 & 1 & 1\end{bmatrix} = \begin{bmatrix}1+0 & 3+0 & 1+5\\ 1+7 & 0+5 & 0+0\\ 1+2 & 2+1 & 2+1\end{bmatrix} = \begin{bmatrix}1 & 3 & 6\\ 8 & 5 & 0\\ 3 & 3 & 3\end{bmatrix}.$$

Another, much less often used notion of matrix addition is the direct sum.

SCALAR MULTIPLICATION

Given a matrix A and a number *c*, the scalar multiplication *cA* is computed by multiplying every element of A by the scalar *c* (i.e.) $(cA)i,j = c \cdot a_{i,j}$). For example:

$$2\cdot\begin{bmatrix}1 & 8 & -3\\ 4 & -2 & 5\end{bmatrix} = \begin{bmatrix}2\cdot 1 & 2\cdot 8 & 2\cdot -3\\ 2\cdot 4 & 2\cdot -2 & 2\cdot 5\end{bmatrix} = \begin{bmatrix}2 & 16 & -6\\ 8 & -4 & 10\end{bmatrix}$$

Matrix addition and scalar multiplication turn the set $M(m, n, \mathbb{R})$ of all *m*-by-*n* matrices with real entries into a real vector space of dimension $m \cdot n$.

MATRIX MULTIPLICATION

Multiplication of two matrices is well-defined only if the number of columns of the left matrix is the same as the number of rows of the right matrix. If A is an *m*-by-*n* matrix and B is an *n*-by-*p* matrix, then their matrix product AB is the *m*-by-*p* matrix given by:

$$(AB)_{i,j} = a_{i,1}b_{1,j} + a_{i,2}b_{2,j} + \dots + a_{i,n}b_{n,j}$$

for each pair (*i,j*). For example:

$$\begin{bmatrix}1 & 0 & 2\\ -1 & 3 & 1\end{bmatrix} \times \begin{bmatrix}3 & 1\\ 2 & 1\\ 1 & 0\end{bmatrix}$$

$$= \begin{bmatrix} (1\times 3+0\times 2+2\times 1) & (1\times 1+0\times 1+2\times 0) \\ (-1\times 3+3\times 2+1\times 1) & (-1\times 1\times 3+1+1\times 0) \end{bmatrix} = \begin{bmatrix} 5 & 1 \\ 4 & 2 \end{bmatrix}.$$

Matrix multiplication has the following properties:

1. (AB)C = A(BC) for all k-by-m matrices A, m-by-n matrices B and n-by-p matrices C ("associativity").

2. (A+B)C = AC+BC for all m-by-n matrices A and B and n-by-k matrices C ("right distributivity").

3. C(A+B) = CA+CB for all m-by-n matrices A and B and k-by-m matrices C ("left distributivity").

Matrix multiplication is not commutative; that is, given matrices A and B and their product defined, then generally AB ≠ BA. It may also happen that AB is defined but BA is not defined.

Besides the ordinary matrix multiplication just described, there exist other operations on matrices that can be considered forms of multiplication, such as the Hadamard product and the Kronecker product.

LINEAR TRANSFORMATIONS

Matrices can conveniently represent linear transformations (also known as "linear maps") between finite-dimensional vector spaces. Let R^n be an n-dimensional vector space, and let the vectors in this space be represented in matrix format as column vectors (n-by-1 matrices). For every linear map f: $R^n \to R^m$ there exists a unique m-by-n matrix A such that

$$f(x) = A_x$$

for each vector x in R^n.

We say that the matrix A "represents" the linear map f, or that A is the "transformation matrix" of f.

Matrix multiplication neatly corresponds to the composition of maps. If the k-by-m matrix B represents another linear map g : $R^m \to R^k$, then the composition $g \circ f$ is represented by BA:

$$(g \circ f)(x) \quad = g(f)(x))\ g(Ax) = B(Ax) = (BA)x.$$

This follows from the above-mentioned associativity of matrix multiplication.

More generally, a linear map from an *n*-dimensional vector space to an *m*-dimensional vector space is represented by an *m*-by-*n* matrix, provided that bases have been chosen for each. This property makes matrices powerful data structures in high-level programming languages.

RANKS

The rank of a matrix A is the dimension of the image of the linear map represented by A; this is the same as the dimension of the space generated by the rows of A, and also the same as the dimension of the space generated by the columns of A. It can also be defined without reference to linear algebra as follows: the rank of an *m*-by-*n* matrix A is the smallest number *k* such that A can be written as a product BC where B is an *m*-by-*k* matrix and C is a *k*-by-*n* matrix (although this is not a practical way to compute the rank).

TRANSPOSE

The transpose of an *m*-by-*n* matrix A is the *n*-by-*m* matrix A^T (also sometimes written as A^{tr}, or ${}^tA$, or A′) formed by turning rows into columns and columns into rows, i.e., $A^T[i, j] = A[j, i]$ for all indices *i* and *j*. If A describes a linear map with respect to two bases, then the matrix A^T describes the transpose of the linear map with respect to the dual bases, see dual space.

We have $(A + B)^T = A^T + B^T$ and $(AB)^T = B^T A^T$.

SQUARE ROOT

Given two bounded matrices T and B, B is a square root of T if T = B*B.

SQUARE MATRICES AND RELATED DEFINITIONS

A square matrix is a matrix which has the same number of rows and columns. The set of all square *n*-by-*n* matrices, together with matrix addition and matrix multiplication is a ring. Unless *n* = 1, this ring is not commutative.

M(n, R), the ring of real square matrices, is a real unitary associative algebra. M(n, C), the ring of complex square matrices, is a complex associative algebra.

The unit matrix or identity matrix I_n of size n is the n-by-n matrix in which all the elements on the main diagonal are equal to 1 and all other elements are equal to 0. The identity matrix is named thus because it satisfies the equations $MI_n = M$ and $I_nN = N$ for any m-by-n matrix M and n-by-k matrix N. For example, if $n = 3$:

$$I_3 = \begin{bmatrix} 1 & 0 & 0 \\ 0 & 1 & 0 \\ 0 & 0 & 1 \end{bmatrix}.$$

The identity matrix is the identity element in the ring of square matrices.

Invertible elements in this ring are called invertible matrices or non-singular matrices. An n-by-n matrix A is invertible if and only if there exists a matrix B such that

$$AB = I_n \ (= BA).$$

In this case, B is the inverse matrix of A, denoted by A^{-1}. The set of all invertible n-by-n matrices forms a group (specifically a Lie group) under matrix multiplication, the general linear group.

If λ is a number and v is a non-zero vector such that $Av = \lambda v$, then we call v an eigenvector of A and λ the associated eigenvalue. (*Eigen* means "own" in German and in Dutch.) The number λ is an eigenvalue of A if and only if $A-\lambda I_n$ is not invertible, which happens if and only if $p_A(\lambda) = 0$, where $p_A(x)$ is the characteristic polynomial of A. $p_A(x)$ is a polynomial of degree n and therefore has n complex roots if multiple roots are counted according to their multiplicity. Every square matrix has at most n complex eigenvalues.

The determinant of a square matrix A is the product of its n eigenvalues, but it can also be defined by the *Leibniz formula.* Invertible matrices are precisely those matrices with a nonzero determinant.

The Gaussian elimination algorithm is of central

importance: it can be used to compute determinants, ranks and inverses of matrices and to solve systems of linear equations.

The trace of a square matrix is the sum of its diagonal entries, which equals the sum of its n eigenvalues.

The matrix exponential function is defined for square matrices, using power series.

VECTOR AND MATRIX

EUCLIDEAN VECTOR AND EUCLIDEAN SPACE

The n–dimensional (real) euclidean space R^n is the collection of n–tuples of numbers, called vectors. Vectors can be either presented horizontally,

$$x = (x_1, x_2, \ldots, x_n)$$

or vertically.

$$x = \begin{bmatrix} x_1 \\ x_2 \\ \cdot \\ \cdot \\ x_n \end{bmatrix}$$

The only thing you need to keep in mind is that, when vectors are mixed with matrices, it is often more convenient to use vertical vectors.

The number n is the dimension of the vector x. The numbers $x_1, x_2, \ldots, x_n$ are called the coordinates of x. Strictly speaking, these are the coordinates with respect to the standard basis vectors

$$e_1 = (1, 0, \ldots, 0),$$
$$e_2 = (0, 1, \ldots, 0),$$
$$\vdots$$
$$e_n = (0, 0, \ldots, 1).$$

We also mention the special zero vector

$$0 = (0, 0, \ldots, 0).$$

Example: The 1–dimensional euclidean space

$R^1 = \{(x): x \text{ is a real number}\}$

can be identified with all real numbers R and be visualized as a straight line.

The 2–dimensional euclidean space
$R^2 = \{(x_1, x_2): x_1, x_2 \text{ are real numbers}\}$
can be visualized as the coordinate plane.

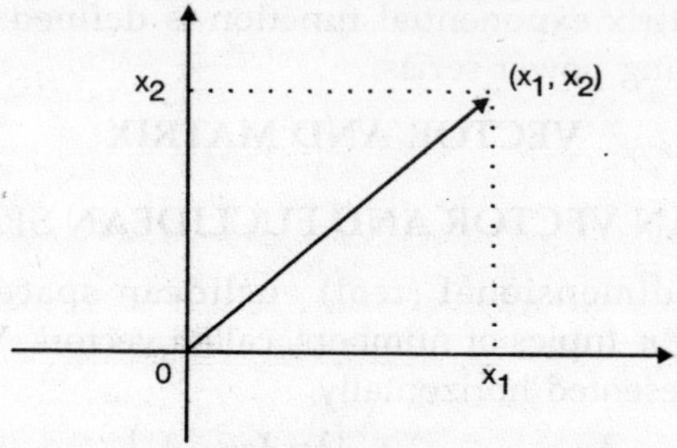

The 3–dimensional euclidean space
$R^3 = \{(x_1, x_2, x_3): x_1, x_2, x_3 \text{ are real numbers}\}$
can be visualized as the world we are living in.

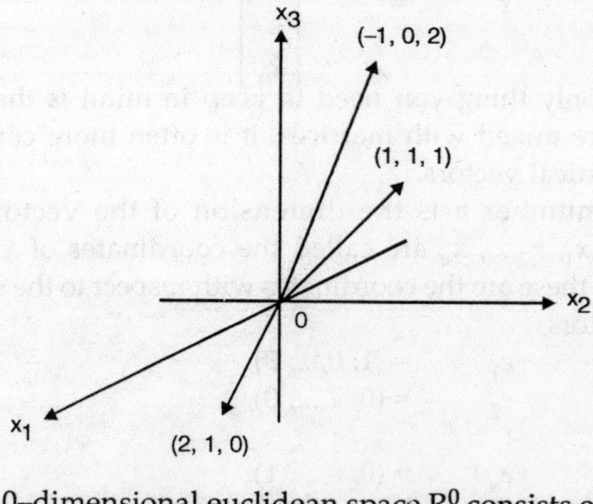

The 0–dimensional euclidean space R^0 consists of a single point – the origin.

The example illustrates that the euclidean vectors can be geometrically presented (in a coordinate system) as an arrow starting from the origin. If an arrow does not start from the origin, we need to parallel shift it so that it begins at the origin.

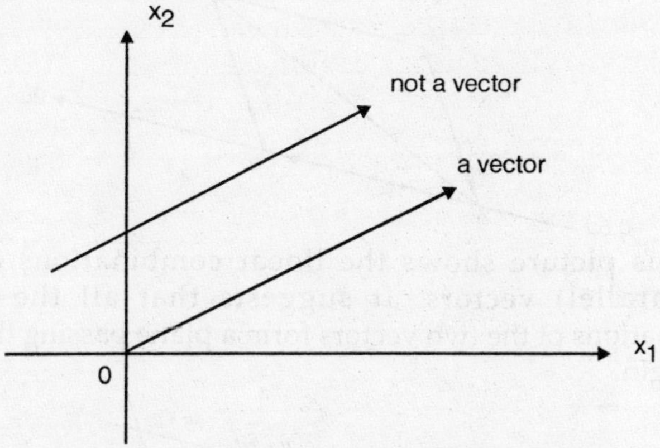

Vectors of the same dimension can be added

$(x_1, x_2, ..., x_n) + (y_1, y_2, ..., y_n) = (x_1 + y_1, x_2 + y_2, ..., x_n + y_n)$.

A number (scalar) can also be multiplied to vectors

$c\,(x_1, x_2, ..., x_n) = (cx_1, cx_2, ..., cx_n)$.

The addition and scalar multiplication can also be combined to form linear combinations $c_1x_1 + c_2x_2 + ... + c_kx_k$.

Example: Let $u = (1, 2)$, $v = (3, 4)$, $w = (5, 6)$, $x = (1, 0, -3)$, $y = (2, -1, 5)$. Then

$$u + v = (1 + 3, 2 + 4) = (4, 6)$$

$$3u = (3\times1, 3\times2) = (3, 6)$$

$$u + v + w = (1 + 3 + 5, 2 + 4 + 6) = (9, 12)$$

$$3u - 4v + w = (3\times1 - 4\times3 + 5, 3\times2 - 4\times4 + 6) = (-4, -4)$$

$$x + y = (1 + 2, 0 - 1, -3 + 5) = (3, -1, 2)$$

$$-2x = (-2\times1, -2\times0, -2\times(-3)) = (-2, 0, 6)$$

$$-2x + 3y = (-2\times1 + 3\times2, -2\times0 + 3\times(-1), -2\times(-3) + 3\times5) = (4, -3, 21)$$

Moreover, because the dimension of u, v, w is different from the dimension of x, y, expressions such as $u + x$, $2v + 3y$ are meaningless.

Geometrically, the addition is described by parallelogram, and the scalar multiplication is the stretching/shrinking and sometimes reversing (for negative scalars) of vectors.

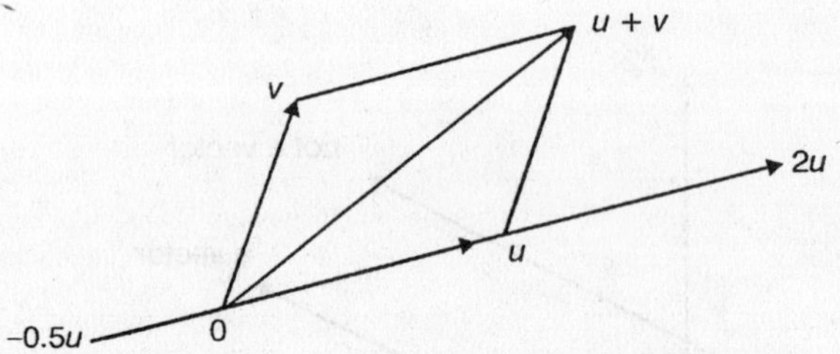

This picture shows the linear combinations of two (nonparallel) vectors. It suggests that all the linear combinations of the two vectors form a plane passing through the origin.

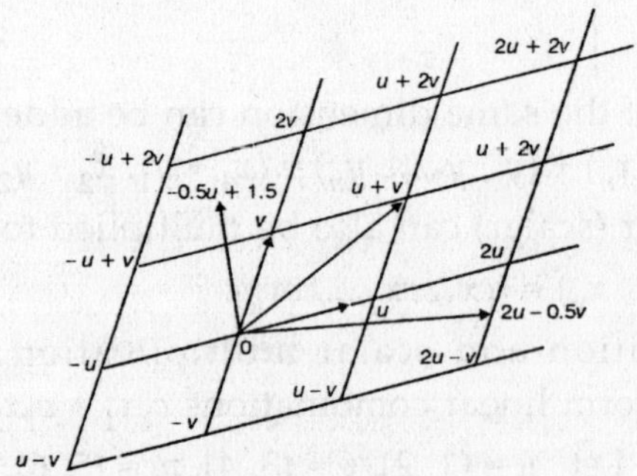

Finally, the operations have the following obvious properties.

commutative

$$u + v \quad = v + u$$

associative

$$(u + v) + w \qquad = u + (v + w)$$

zero

$$u + 0 \quad = u = 0 + u$$

distributive

$$r(u + v) = ru + rv$$

distributive

$$(r + s)u = ru + su$$

associative

$$(rs)u \quad = r(su)$$

VECTOR ADDITION IS GIVEN BY PARALLELOGRAM

Let us look at the addition of two vectors (red and blue in the picture) in R^2. The green vector is constructed from the parallelogram.

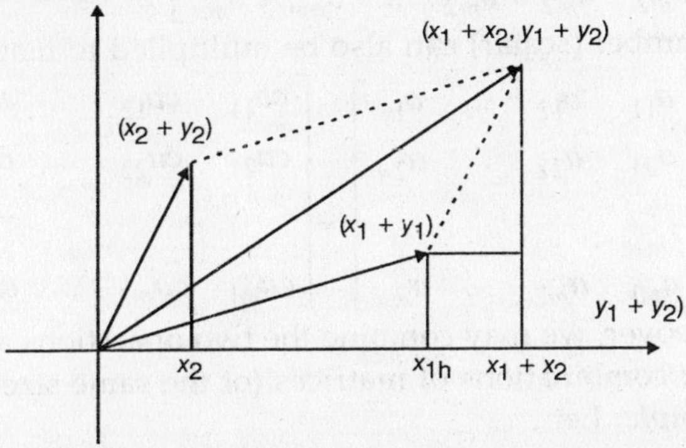

The parallelogram tells us that the two right triangles (in gray shade) are identical. In particular, the two pink segments of the x–axis have the same length. This implies that the x–coordinate of the green vector is $x_1 + x_2$. By similar argument, the y–coordinate of the green vector is $y_1 + y_2$.

MATRIX

Let $M(m, n)$ be the collection of all m by n matrices. The zero matrix is

$$O = \begin{bmatrix} 0 & 0 & \dots & 0 \\ 0 & 0 & \dots & 0 \\ \vdots & \vdots & & \vdots \\ 0 & 0 & \dots & 0 \end{bmatrix}$$

Like vectors, matrices of the same size may be added.

$$\begin{bmatrix} a_{11} & a_{12} & \dots & a_{1n} \\ a_{21} & a_{22} & \dots & a_{2n} \\ \vdots & \vdots & & \vdots \\ a_{m1} & a_{m2} & \dots & a_{mn} \end{bmatrix} + \begin{bmatrix} b_{11} & b_{12} & \dots & b_{1n} \\ b_{21} & b_{22} & \dots & b_{2n} \\ \vdots & \vdots & & \vdots \\ b_{m1} & b_{m2} & \dots & b_{mn} \end{bmatrix}$$

$$= \begin{bmatrix} a_{11} + b_{11} & a_{12} + b_{12} & \dots & a_{1n} + b_{1n} \\ a_{21} + b_{21} & a_{22} + b_{22} & \dots & a_{2n} + b_{2n} \\ \vdots & \vdots & & \vdots \\ a_{m1} + b_{m2} & a_{m2} + b_{m2} & \dots & a_{mn} + b_{mn} \end{bmatrix}$$

A number (scalar) can also be multiplied to matrices

$$c \begin{bmatrix} a_{11} & a_{12} & \dots & a_{1n} \\ a_{21} & a_{22} & \dots & a_{2n} \\ \vdots & \vdots & & \vdots \\ a_{m1} & a_{m2} & \dots & a_{mn} \end{bmatrix} = \begin{bmatrix} ca_{11} & ca_{12} & \dots & ca_{1n} \\ ca_{21} & ca_{22} & \dots & ca_{2n} \\ \vdots & \vdots & & \vdots \\ ca_{m1} & ca_{m2} & \dots & ca_{mn} \end{bmatrix}$$

Moreover, we may combine the two operations and form the linear combinations of matrices (of the same size).

***Example*:** Let

$$A = \begin{bmatrix} 1 & 1 & -2 \\ 3 & 0 & 1 \end{bmatrix}, B = \begin{bmatrix} -1 & 0 & 1 \\ -2 & 1 & 1 \end{bmatrix}, C = \begin{bmatrix} 0 & 1 & 0 \\ 1 & 0 & 0 \end{bmatrix}, D = \begin{bmatrix} 0 & 1 \\ 1 & 2 \end{bmatrix}$$

Then

$$A + B = \begin{bmatrix} 1+(-1) & 1+0 & (-2)+1 \\ 3+(-2) & 0+1 & 1+1 \end{bmatrix} = \begin{bmatrix} 0 & 1 & -1 \\ 1 & 1 & 2 \end{bmatrix}$$

$$A + B + C = \begin{bmatrix} 1+(-1)+0 & 1+0+1 & (-2)+1+0 \\ 3+(-2)+1 & 0+1+0 & 1+1+0 \end{bmatrix}$$

$$= \begin{bmatrix} 0 & 2 & -1 \\ 2 & 1 & 2 \end{bmatrix}$$

$$2A = \begin{bmatrix} 2\times 1 & 2\times 1 & 2\times(-2) \\ 2\times 3 & 2\times 0 & 2\times 1 \end{bmatrix} = \begin{bmatrix} 2 & 2 & -4 \\ 6 & 0 & 2 \end{bmatrix}$$

$$-3D = \begin{bmatrix} (-3)\times 0 & (-3)\times 1 \\ (-3)\times 1 & (-3)\times 2 \end{bmatrix} = \begin{bmatrix} 0 & -3 \\ -3 & -6 \end{bmatrix}$$

$$3A - 4B + 2C = \begin{bmatrix} 3+4+0 & 3+0+2 & -6-4+0 \\ 9+8+2 & 0-4+0 & 3-4+0 \end{bmatrix}$$

$$= \begin{bmatrix} 7 & 5 & -10 \\ 19 & -4 & -1 \end{bmatrix}$$

On the other hand, expressions such as $A + D$, $B + D$, $A + D$ are meaningless.

Euclidean vectors can be considered as matrices of special sizes. Specifically, a vertical n–dimensional euclidean vector is an n by 1 matrix, and the euclidean space R^n is identified with $M(n, 1)$. On the other hand, if we present vectors horizontally, then R^n is identified with $M(1, n)$.

Although euclidean vectors may be presented either vertically or horizontally, you are strongly advised to use vertical vectors when they are mixed with matrices. For example, in expressing a system of linear equation as $Ax = b$, because of the presence of the matrix A, we should consider x and b as vertical vectors.

Conversely, matrices can be considered as an (ordered) collection of vectors. Specifically, an m by n matrix is equivalent to n euclidean column vectors of dimension m.

$$A = [a_1\ a_2 \ldots\ a_n] \Leftrightarrow a_1, a_2, \ldots, a_n \in R^n.$$

Similarly, we may also consider row vectors, so that an m by n matrix is equivalent to m euclidean row vectors of dimension n.

Example: The matrix

$$A = \begin{bmatrix} 1 & 2 & 3 \\ 4 & 5 & 6 \end{bmatrix}$$

can be considered as three 2–dimensional column vectors

$$A = [a_1\ a_2\ a_3],\ a_1 = \begin{bmatrix} 1 \\ 4 \end{bmatrix},\ a_2 = \begin{bmatrix} 2 \\ 5 \end{bmatrix},\ a_3 = \begin{bmatrix} 3 \\ 6 \end{bmatrix}$$

It can also be considered as two 3–dimensional row vectors.

$$A=\begin{bmatrix} b_1 \\ b_2 \end{bmatrix}, b_1 = (1, 2, 3), b_2 = (4, 5, 6)$$

The relations between the vectors and the matrices are compatible with addition and scalar multiplication. For example, for three column matrices $A = [a_1\ a_2\ a_3]$, $B = [b_1\ b_2\ b_3]$, we have

$$A + B= [a_1 + b_1\ a_2 + b_2\ a_3 + b_3],\ cA = [ca_1\ ca_2\ ca_3].$$

Because of this compatibility, the properties for the addition and scalar multiplication of vectors are also hold for matrices.

SOME MATRIX NOTATIONS AND TERMINOLOGIES

In theoretical discussions about matrices, we sometimes use the notation $A = (a_{ij})$ to denote a matrix, where a_{ij} is the entry at the i–th row and j–th column (the ij–entry). Sometimes we even indicate the size by writing $A = (a_{ij})_{m\times n}$. In terms of this compact notation, we have

$$(a_{ij}) + (b_{ij}) = (a_{ij} + b_{ij}),\ c(a_{ij}) = (ca_{ij}).$$

Example:

$$(i + j)_{3\times3}=\begin{bmatrix} 1+1 & 1+2 & 1+3 \\ 2+1 & 2+2 & 2+3 \\ 3+1 & 3+2 & 3+3 \end{bmatrix} = \begin{bmatrix} 2 & 3 & 4 \\ 3 & 4 & 5 \\ 4 & 5 & 6 \end{bmatrix}$$

$$(ij)_{4\times5} = \begin{bmatrix} 1\times1 & 1\times2 & 1\times3 & 1\times4 & 1\times5 \\ 2\times1 & 2\times2 & 2\times3 & 2\times4 & 2\times5 \\ 3\times1 & 3\times2 & 3\times3 & 3\times4 & 3\times5 \\ 4\times1 & 4\times2 & 4\times3 & 4\times4 & 4\times5 \end{bmatrix} = \begin{bmatrix} 1 & 2 & 3 & 4 & 5 \\ 2 & 4 & 6 & 8 & 10 \\ 3 & 6 & 9 & 12 & 15 \\ 4 & 8 & 12 & 16 & 20 \end{bmatrix}$$

$$(j^2)_{4\times3}=\begin{bmatrix} 1^2 & 2^2 & 3^2 \\ 1^2 & 2^2 & 3^2 \\ 1^2 & 2^2 & 3^2 \\ 1^2 & 2^2 & 3^2 \end{bmatrix} = \begin{bmatrix} 1 & 4 & 9 \\ 1 & 4 & 9 \\ 1 & 4 & 9 \\ 1 & 4 & 9 \end{bmatrix}$$

For an n by n square matrix $A = (a_{ij})$, the terms $a_{11}, a_{22}, \ldots, a_{nn}$ are the diagonal entries of the matrix. A is a diagonal matrix if all the off–diagonal entries are zero. In other words, $a_{ij} = 0$ for $i \neq j$. The following are all the 2 by 2 and 3 by 3 diagonal matrices.

$$\begin{bmatrix} a & 0 \\ 0 & b \end{bmatrix}$$

$$\begin{bmatrix} a & 0 & 0 \\ 0 & b & 0 \\ 0 & 0 & c \end{bmatrix}$$

A is an upper triangular matrix if all the entries below the diagonal are zero. In other words, $a_{ij} = 0$ for $i > j$. The following are all the 2 by 2 and 3 by 3 upper triangular matrices.

$$\begin{bmatrix} a & b \\ 0 & c \end{bmatrix}$$

$$\begin{bmatrix} a & b & c \\ 0 & d & e \\ 0 & 0 & f \end{bmatrix}$$

Similarly, A is a lower triangular matrix if all the entries above the diagonal are zero. In other words, $a_{ij} = 0$ for $i < j$. The following are all the 2 by 2 and 3 by 3 upper triangular matrices.

$$\begin{bmatrix} a & 0 \\ b & c \end{bmatrix}$$

$$\begin{bmatrix} a & 0 & 0 \\ b & c & 0 \\ d & e & f \end{bmatrix}$$

Clearly, A is upper and lower triangular at the same time if and only if it is diagonal.

Exercise: Draw the vectors (1, 3), (0, 1), (0, 0), (–1, –3), (2, –2).

Answer:

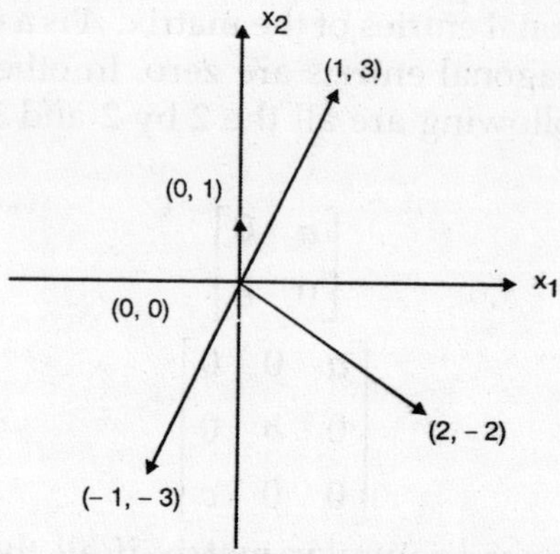

Exercise: Draw the collections of vectors.

$X = \{(0, y): y \text{ is any number}\}$
$Y = \{(x, 0): x \text{ is any number}\}$
$S = \{(x, y): x^2 + y^2 = 1\}$
$D = \{(x, x): x \text{ is any number}\}$
$E = \{(x, y): x + 3y = 0\}$

Answer:

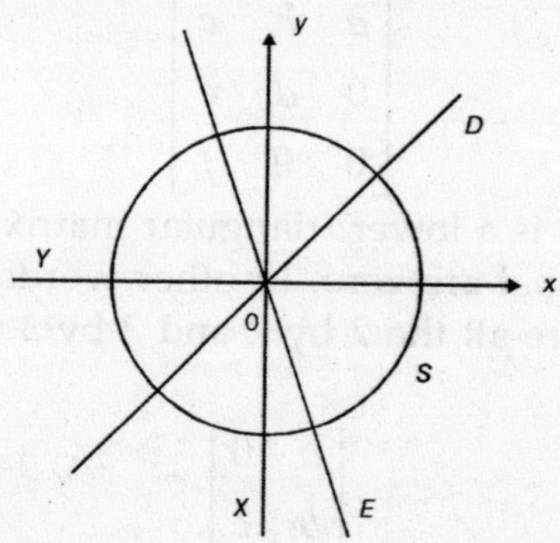

Exercise: Compute.

Answer:

$$\begin{bmatrix}1\\3\end{bmatrix}-\begin{bmatrix}5\\6\end{bmatrix}=\begin{bmatrix}-4\\-3\end{bmatrix},\begin{bmatrix}2\\4\end{bmatrix}-\begin{bmatrix}7\\8\end{bmatrix}=\begin{bmatrix}-5\\-4\end{bmatrix},3\begin{bmatrix}1\\3\end{bmatrix}-5\begin{bmatrix}5\\6\end{bmatrix}$$

$$=\begin{bmatrix}-22\\-21\end{bmatrix},3\begin{bmatrix}2\\4\end{bmatrix}-5\begin{bmatrix}7\\8\end{bmatrix}=\begin{bmatrix}-29\\-28\end{bmatrix}$$

$$\begin{bmatrix} 1 & 2 \\ 3 & 4 \end{bmatrix} - \begin{bmatrix} 5 & 7 \\ 6 & 8 \end{bmatrix} = \begin{bmatrix} -4 & -5 \\ -3 & -4 \end{bmatrix}, 3\begin{bmatrix} 1 & 2 \\ 3 & 4 \end{bmatrix} - 5\begin{bmatrix} 5 & 7 \\ 6 & 8 \end{bmatrix}$$

$$\begin{bmatrix} 0 & -2 & 3 \\ 1 & 0 & -3 \\ 2 & -2 & 2 \end{bmatrix} + \begin{bmatrix} 2 & 0 & -1 \\ 0 & 4 & -3 \\ 3 & -3 & 0 \end{bmatrix} = \begin{bmatrix} 2 & -2 & 2 \\ 1 & 4 & -6 \\ 5 & -5 & 2 \end{bmatrix},$$

$$-2\begin{bmatrix} 0 & -2 & 3 \\ 1 & 0 & -3 \\ 2 & -2 & 2 \end{bmatrix} + 3\begin{bmatrix} 2 & 0 & -1 \\ 0 & 4 & -3 \\ 3 & -3 & 0 \end{bmatrix} = \begin{bmatrix} 4 & 4 & -9 \\ -2 & 12 & -15 \\ 5 & -5 & -4 \end{bmatrix}$$

$$a\begin{bmatrix} 1 & 0 & 0 \\ 0 & 0 & 0 \\ 0 & 0 & 0 \end{bmatrix} + b\begin{bmatrix} 0 & 0 & 0 \\ 0 & 1 & 0 \\ 0 & 0 & 0 \end{bmatrix} + c\begin{bmatrix} 0 & 0 & 0 \\ 0 & 0 & 0 \\ 0 & 0 & 1 \end{bmatrix} = \begin{bmatrix} a & 0 & 0 \\ 0 & b & 0 \\ 0 & 0 & c \end{bmatrix}$$

$$a_{11}\begin{bmatrix} 1 & 0 \\ 0 & 0 \end{bmatrix} + a_{12}\begin{bmatrix} 0 & 1 \\ 0 & 0 \end{bmatrix} + a_{21}\begin{bmatrix} 0 & 0 \\ 1 & 0 \end{bmatrix} + a_{22}\begin{bmatrix} 0 & 0 \\ 0 & 1 \end{bmatrix} = \begin{bmatrix} a_{11} & a_{12} \\ a_{21} & a_{22} \end{bmatrix}$$

$$\begin{bmatrix} 1 & -3 & 5 & 10 \\ 0 & 5 & 1 & -7 \\ 1 & 0 & 9 & -3 \\ 2 & 0 & 1 & 0 \\ -6 & 3 & 0 & 0 \end{bmatrix} + \begin{bmatrix} 4 & -9 & 8 \\ 4 & 0 & 3 \\ 1 & 4 & 0 \\ 0 & 0 & 4 \\ 0 & 5 & 3 \end{bmatrix} = 12\begin{bmatrix} 0 & 0 & 0 & 0 & 0 \\ 0 & 0 & 0 & 0 & 0 \\ 0 & 0 & 0 & 0 & 0 \end{bmatrix}$$

$$-13\begin{bmatrix} 0 & 0 & 0 & 0 & 0 \\ 0 & 0 & 0 & 0 & 0 \\ 0 & 0 & 0 & 0 & 0 \\ 0 & 0 & 0 & 0 & 0 \\ 0 & 0 & 0 & 0 & 0 \end{bmatrix}$$

The last two are meaningless because the sizes do not match.

Exercise: Take the standard basis vectors $e_1 = (1, 0, 0)$, $e_2 = (0, 1, 0)$, $e_3 = (0, 0, 1)$ of R^3. Compute the linear combination $x_1e_1 + x_2e_2 + x_3e_3$. Then generalize your result to other dimensions.

Answer: $x_1e_1 + x_2e_2 + x_3e_3 = x_1(1, 0, 0) + x_2(0, 1, 0) + x_3(0, 0, 1) = (x_1, 0, 0) + (0, x_2, 0) + (0, 0, x_3) = (x_1, x_2, x_3)$. In general, for the standard basis $e_1 = (1, 0,..., 0)$, $e_2 = (0, 1,..., 0)$,..., $e_n = (0, 0,..., 1)$ of R^n, we have $(x_1, x_2,..., x_n) = x_1e_1 + x_2e_2 +... + x_ne_n$.

Exercise: For $u = (1, 2)$, $v = (3, 4)$, $w = (5, 6)$, compute the linear combination $x_1u + x_2v + x_3w$. Do the same for $a_1 = (1, 2, 3)$, $a_2 = (4, 5, 6)$, $a_3 = (7, 8, 9)$. Express your computation in vertical vectors and draw a general observation.

Answer:

$$x_1\begin{bmatrix}1\\2\end{bmatrix}+x_2\begin{bmatrix}3\\4\end{bmatrix}+x_3\begin{bmatrix}5\\6\end{bmatrix}$$

$$=\begin{bmatrix}x_1\\2x_1\end{bmatrix}+\begin{bmatrix}3x_2\\4x_2\end{bmatrix}+\begin{bmatrix}5x_3\\6x_3\end{bmatrix}=\begin{bmatrix}x_1+3x_2+5x_3\\2x_1+4x_2+6x_3\end{bmatrix}$$

$$x_1\begin{bmatrix}1\\2\\3\end{bmatrix}+x_2\begin{bmatrix}4\\5\\6\end{bmatrix}+x_3\begin{bmatrix}7\\8\\9\end{bmatrix}$$

$$=\begin{bmatrix}x_1\\2x_1\\3x_1\end{bmatrix}+\begin{bmatrix}4x_2\\5x_2\\6x_2\end{bmatrix}+\begin{bmatrix}7x_3\\8x_3\\9x_3\end{bmatrix}=\begin{bmatrix}x_1+4x_2+7x_3\\2x_1+5x_2+8x_3\\3x_1+6x_2+9x_3\end{bmatrix}$$

The results are of the form Ax, with the matrix A being respectively

$$\begin{bmatrix}1&3&5\\2&4&6\end{bmatrix}=[u\ v\ w]\quad\begin{bmatrix}1&4&7\\2&5&8\\3&6&9\end{bmatrix}=[a_1\ a_2\ a_3]$$

In general, we have

$x_1a_1 + x_2a_2 +... + x_na_n = Ax$

where $A = [a_1\ a_2... a_n]$ is a matrix with the vectors $a_1, a_2,..., a_n$ as columns.

Exercise: Express the vectors $b = (4, 6)$, $c = (1, 1)$, $e_1 = (1, 0)$, $e_2 = (0, 1)$ as linear combinations of the vectors $u = (1, 2)$, $v = (3, 4)$, $w = (5, 6)$. Can you express as linear combinations of the vectors u, v only?

Answer: We would like to write b, c, e_1, e_2 in the form $x_1(1, 2) + x_2(3, 4) + x_3(5, 6)$. From the previous exercise, we have

$$x_1\begin{bmatrix}1\\2\end{bmatrix} + x_2\begin{bmatrix}3\\4\end{bmatrix} + x_3\begin{bmatrix}5\\6\end{bmatrix} = \begin{bmatrix}x_1 + 3x_2 + 5x_3\\2x_1 + 4x_2 + 6x_3\end{bmatrix}$$

Therefore the problem is to solve the system of equations

$$x_1 + 3x_2 + 5x_3 = b_1$$
$$2x_1 + 4x_2 + 6x_3 = b_2$$

For $b_1 = 4$, $b_2 = 6$ on the right side, we find $x_1 = 1$, $x_2 = 1$ is one solution. Therefore $b = u + v$. In fact, the solution is not unique. For example, $x_1 = 2$, $x_2 = -1$, $x_3 = 1$ is another solution, from which we get $b = 2u - v + w$.

Incidentally, we have already expressed b as a linear combination of u and v: $b = u + v$. Such expressions are equivalent to solutions of

$$x_1 + 3x_2 = b_1$$
$$2x_1 + 4x_2 = b_2$$

We do the similar computation for the other vectors and find $c = -1/2\,u + 1/2\,v$, $e_1 = -2u + v$, $e_2 = 3/2\,u - 1/2\,v$.

Exercise: Express the vectors $b = (1, 1, 1)$, $c = (-1, 0, 1)$, $e_1 = (1, 0, 0)$ as linear combinations of the vectors $a_1 = (1, 2, 3)$, $a_2 = (4, 5, 6)$, $a_3 = (7, 8, 9)$. Is the expression (i.e., the coefficients) unique?

Answer: We would like to write b, c, e_1 in the form $x_1(1, 2, 3) + x_2(4, 5, 6) + x_3(7, 8, 9)$. The computation is similar to the previous exercise. From this exercise, we have

$$x_1\begin{bmatrix}1\\2\\3\end{bmatrix} + x_2\begin{bmatrix}4\\5\\6\end{bmatrix} + x_3\begin{bmatrix}7\\8\\9\end{bmatrix} = \begin{bmatrix}x_1 + 4x_2 + 7x_3\\2x_1 + 5x_2 + 8x_3\\3x_1 + 6x_2 + 9x_3\end{bmatrix}$$

Therefore the problem is to solve the system.

$$x_1 + 4x_2 + 7x_3 = b_1$$
$$2x_1 + 5x_2 + 8x_3 = b_2$$

$$3x_1 + 6x_2 + 9x_3 = b_3$$

For $b_1 = b_2 = b_3 = 1$ on the right side, we find $x_1 = -1/3$, $x_2 = 1/3$, $x_3 = 0$ is one solution. Therefore $b = -1/3\ a_1 + 1/3\ a_2$. Since the solution is not unique, the expression is not unique. Another possibility is $b = -1/6\ a_1 + 1/6\ a_3$.

For $b_1 = -1$, $b_2 = 0$, $b_3 = 1$ on the right side, we find $x_1 = 4/3$, $x_2 = 0$, $x_3 = -1/3$ is one solution. Therefore $c = 4/3\ a_1 - 1/3\ a_3$. Again the expression is not unique.

For $b_1 = 1$, $b_2 = 0$, $b_3 = 0$ on the right side, there is no solution. Therefore e_1 is not a linear combination of a_1, a_2, a_3.

Exercise: Express the matrix

$$X = \begin{bmatrix} 3 & 4 \\ 2 & 3 \end{bmatrix}$$

as linear combinations of the matrices.

$$A = \begin{bmatrix} 1 & 2 \\ 0 & 1 \end{bmatrix}, B = \begin{bmatrix} 1 & 0 \\ 2 & 1 \end{bmatrix}, C = \begin{bmatrix} 0 & 1 \\ -1 & 0 \end{bmatrix}$$

Is the expression (i.e., the coefficients) unique? Is it true that any 2 by 2 matrix is a linear combination of A, B and C?

Answer: By comparing the four entries on the both side cf the equation $x_1A + x_2B + x_3C = X$,

$$x_1 \begin{bmatrix} 1 & 2 \\ 0 & 1 \end{bmatrix} + x_2 \begin{bmatrix} 1 & 0 \\ 2 & 1 \end{bmatrix} + x_3 \begin{bmatrix} 0 & 1 \\ -1 & 0 \end{bmatrix} = \begin{bmatrix} 3 & 4 \\ 2 & 3 \end{bmatrix}$$

we have the following system of equations.

$$\begin{aligned} x_1 + x_2 &= 3 \\ 2x_1 + x_3 &= 4 \\ 2x_2 - x_3 &= 2 \\ x_1 + x_2 &= 3 \end{aligned}$$

From the row echelon form of the augmented matrix,

$$\left[\begin{array}{ccc|c} 1 & 1 & 0 & 3 \\ 0 & 2 & -1 & 2 \\ 0 & 0 & 0 & 0 \\ 0 & 0 & 0 & 0 \end{array}\right]$$

we see that the system has non–unique solutions. One solution is $x_1 = 2, x_2 = 1, x_3 = 0$. Therefore $X = 2A + 1B + 0C$. Moreover, the expression is not unique.

For a general 2 by 2 matrix $X = (b_{ij})$ (i.e., b_{ij} is the entry of X at the i–th row and j–th column), the corresponding system is

$$\begin{aligned} x_1 + x_2 &= b_{11} \\ 2x_1 + x_3 &= b_{12} \\ 2x_2 - x_3 &= b_{21} \\ x_1 + x_2 &= b_{22} \end{aligned}$$

Then from the row echelon form above (or by the fact that there are more equations than variables), we see that solutions do not always exist. In other words, not any X can be expressed as a linear combination of A, B and C.

Remark: In an earlier exercise, we have seen how the problem of expressing a vector as a linear combination of some other vectors is equivalent to solving a system of linear equations. This exercise shows that we have similar relation for matrices.

LINEAR TRANSFORMATION

DEFINITION

Transformations are rules that convert one data into another data. We may denote a transformation by $T: X \to Y$, where T is the transformation (also called map, function), X is the source set (or domain), and Y is the target set (or codomain). Taking any element $x \in X$ as input, the transformation produces an element $y = T(x) \in Y$ as output. We also call y the image (or value) of x under the transformation T. We may also include more specific information about the transformation by writing $T: X \to Y, x \to$

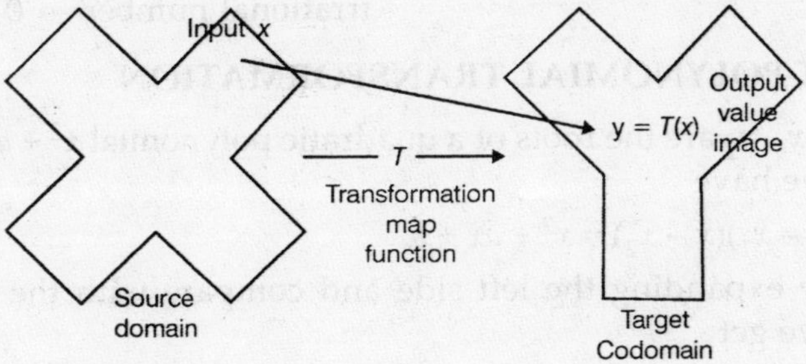

A more rigorous theory of transformations can be found here.

Example: The following are some transformations in everyday life.

Age: people → number

ID_s: student → number

ID_p: professor → number

Instructor: course → professor

Maker: product → manufacturer

Capital City: country → city

Population: city → number

Example In mathematics, transformations are often given by formulae.

Square: $R \to R$, $x \to x^2$

Reflection in x–axis: $R^2 \to R^2$, $(x, y) \to (x, -y)$

Polar–Cartesian: $R^2 \to R^2$, $(r, è) \to (r\cos è, r\sin è)$

Root–Polynomial: $R^2 \to R^2$, $(x_1, x_2) \to (x_1 + x_2, x_1x_2)$

T: $R^3 \to R^2$, $(x_1, x_2, x_3) \to (x_1 + 2x_2 + 3x_3, 4x_1 + 5x_2 + 6x_3)$

Here are two examples of mathematical transformations given by descriptions instead of formulae.

Sign: R → R, positive number → 1, negative number → –1, 0 → 0

Dirichlet function: R → R, rational number → 1, irrational number → 0

ROOT-POLYNOMIAL TRANSFORMATION

If x_1, x_2 are the roots of a quadratic polynomial $x^2 + ax + b$, then we have

$$(x - x_1)(x - x_2) = x^2 + ax + b.$$

By expanding the left side and compare with the right side, we get

$$a = -(x_1 + x_2),\ b = x_1 x_2.$$

Since the polynomial is specified by the two coefficients a and b, we see that the transformation roots $(x_1, x_2) \rightarrow$ polynomial $x^2 + ax + b$ is essentially the same as

$$R^2 \rightarrow R^2,\ (x_1, x_2) \rightarrow (x_1 + x_2,\ x_1 x_2)$$

As an exercise for yourself, how would you define the transformation *Root–Polynomial*: $R^3 \rightarrow R^3$ for three roots of polynomials of degree 3?

Finally, you may try the general case *Root–Polynomial*: $R^n \rightarrow R^n$.

MATRIX TRANSFORMATION

We may think of the left side of a system of linear equations as a transformation. For example, the left side of the system

$$\begin{aligned} x_1 + 2x_2 + 3x_2 &= b_1 \\ 4x_1 + 5x_2 + 6x_4 &= b_2 \end{aligned}$$

can be considered as the linear transformation T in the example above. By expressing vectors in vertical form, we also see that

$$T\begin{bmatrix} x_1 \\ x_2 \\ x_3 \end{bmatrix} = \begin{bmatrix} x_1 + 2x_2 + 3x_3 \\ 4x_1 + 5x_2 + 6x_3 \end{bmatrix} = \begin{bmatrix} 1 & 2 & 3 \\ 4 & 5 & 6 \end{bmatrix}\begin{bmatrix} x_1 \\ x_2 \\ x_3 \end{bmatrix}$$

In particular, the formula for the transformation T is given by a matrix A:

$$T(x) = Ax.$$

In general, for any m by n matrix A, we may construct a transformation T: $R^n \rightarrow R^m$ by the above formula. Such transformations are called matrix transformations.

Example: The following matrices

$$\begin{bmatrix} 1 & -1 & 1 \\ 3 & 1 & -1 \\ 2 & 2 & 1 \end{bmatrix}$$

$$\begin{bmatrix} 1 & 3 & 2 & 0 & 1 \\ -1 & -1 & -1 & 1 & 0 \\ 0 & 4 & 2 & 4 & 3 \\ 1 & 3 & 2 & -2 & 0 \end{bmatrix}$$

give us matrix transformations,

$$T_1 \begin{bmatrix} x_1 \\ x_2 \\ x_3 \end{bmatrix} = \begin{bmatrix} x_1 - x_2 + x_3 \\ 3x_1 + x_2 - x_3 \\ 2x_1 + x_2 + x_3 \end{bmatrix} : R^3 \to R^3$$

$$T_1 \begin{bmatrix} x_1 \\ x_2 \\ x_3 \\ x_4 \\ x_5 \\ x_6 \end{bmatrix} = \begin{bmatrix} x_1 + 3x_2 + 2x_3 + x_5 \\ -x_1 - x_2 - x_3 + x_4 + x_6 \\ 4x_2 + 2x_3 + 4x_4 + 3x_5 + 3x_6 \\ x_1 + 3x_2 + 2x_3 - 2x_4 \end{bmatrix} : R^6 \to R^4$$

or written horizontally,

T_1: $R^3 \to R^3, (x_1, x_2, x_3) \to (x_1 - x_2 + x_3, 3x_1 + x_2 - x_3, 2x_1 + 2x_2 + x_3)$

T_2: $R^6 \to R^4, (x_1, x_2, x_3, x_4, x_5) \to (x_1 + 3x_2 + 2x_3 + x_5, -x_1 - x_2 - x_3 + x_4 + x_6, 4x_2 + 2x_3 + 4x_4 + 3x_5 + 3x_6, x_1 + 3x_2 + 2x_3 - 2x_4)$

Example The transformations

S: $R^3 \to R^3, (x_1, x_2, x_3) \to (x_1 - x_2 + x_3, 3x_1 + x_2 - x_3 - 2, 2x_1 + 2x_2 + x_3)$

$$T: R^2 \to R^2, (x, y) \to (xy, x^2)$$

are not matrix transformations because they cannot be expressed as Ax for a *constant* matrix A.

LINEAR TRANSFORMATION

The following example should convince you that the expression Ax is linear:

$A(x + y) = Ax + Ay$, $A(cx) = cAx$.

A more general proof can be found here.

Example: Let

$$A = \begin{bmatrix} a & b \\ c & d \end{bmatrix}$$

The corresponding linear transformation is

$$T\begin{bmatrix} x_1 \\ x_2 \end{bmatrix} = \begin{bmatrix} ax_1 + bx_2 \\ cx_1 + dx_2 \end{bmatrix} : R^2 \to R^2$$

We have

$$T\begin{bmatrix} x_1 + y_1 \\ x_2 + y_2 \end{bmatrix} = \begin{bmatrix} a(x_1 + y_1) + b(x_2 + y_2) \\ c(x_1 + y_1) + d(x_2 + y_2) \end{bmatrix}$$

$$= \begin{bmatrix} ax_1 + bx_2 \\ cx_1 + dx_2 \end{bmatrix} + \begin{bmatrix} ay_1 + by_2 \\ cy_1 + dy_2 \end{bmatrix} = T\begin{bmatrix} x_1 \\ x_2 \end{bmatrix} + T\begin{bmatrix} y_1 \\ y_2 \end{bmatrix}$$

and (we use k instead of c as the scalar to avoid confusion in notations)

$$T\begin{bmatrix} kx_1 \\ kx_2 \end{bmatrix} = \begin{bmatrix} akx_1 + bkx_2 \\ ckx_2 + dkx_2 \end{bmatrix} = \begin{bmatrix} k(ax_1 + bkx_2) \\ k(cx_1 + dx_2) \end{bmatrix}$$

$$= k\begin{bmatrix} ay_1 + by_2 \\ cy_1 + dy_2 \end{bmatrix} kT\begin{bmatrix} x_1 \\ x_2 \end{bmatrix}$$

You are strongly encouraged to verify the linearity for matrices of other sizes.

The discussion suggests us to introduce the following concept.

A transformation $T: R^n \to R^m$ is a linear transformation if it satisfies

$$T(u + v) = T(u) + T(v),\ T(cu) = cT(u).$$

By taking $c = 0$ in the definition, we have

$$T(0) = 0.$$

Moreover, by combining the two properties in the definition, we see that linear transformations preserve linear combinations.

$$T(c_1u_1 + c_2u_2 + ... + c_ku_k) = c_1T(u_1) + c_2T(u_2) + ... + c_kT(u_k).$$

Example Suppose $T: R^2 \to R^2$ is a linear transformation. Suppose we also know (recall $e_1 = (1, 0)$, $e_2 = (0, 1)$)

$$T(e_1) = (1, 2),\ T(e_2) = (3, 4)$$

Then by

$$(5, -3) = 5\,(1, 0) - 3(0,1) = 5e_1 - 3e_2,$$
$$(-2, 4) = -2e_1 + 4e_2,$$

we have

$$T(5, -3) = 5T(e_1) - 3T(e_2) = 5(1, 2) - 3(3, 4) = (-4, -2),$$
$$T(-2, 4) = -2T(e_1) + 4T(e_2) = -2(1, 2) + 4(3, 4) = (10, 12).$$

In general, by

$$(x_1, x_2) = x_1e_1 + x_2e_2,$$

we have

$T(x_1, x_2) = x_1T(e_1) + x_2T(e_2) = x_1(1, 2) + x_2(3, 4) = (x_1 + 3x_2, 2x_1 + 4x_2)$.

We remark that T is a matrix transformation given by

$$\begin{bmatrix} 1 & 3 \\ 2 & 4 \end{bmatrix} = [T(e_1)\ T(e_2)]$$

Example: The transformation

$S(x_1, x_2, x_3) = (2x_1 + 3x_2 - 7x_3 + 1, 10x_1 - 4x_2 - 8x_3, -4x_1 + x_2 - 5)$

is not a linear transformation because

$$S(0) = S(0, 0, 0) = (1, 0, -5) \text{ ‘‘’ } 0$$

The transformation

$$T(x, y) = (x + y, xy)$$

does satisfy $T(0) = 0$. However, we have

$T(1, 0) + T(0, 1) = (1 + 0, 1\times0) + (0 + 1, 0\times1) = (1, 0) + (1, 0) = (2, 0)$.

This is different from

$T((1, 0) + (0, 1)) = T(1, 1) = (1 + 1, 1\times1) = (2, 1)$.

Therefore T is not linear.

Since the transformations are not linear, they are not matrix transformations.

The linearity of Ax helps us to construct solutions of systems of linear equations. The key idea is that, if x and x' are respectively solutions of $Ax = b$ and $Ax = b'$, then $cx + c'x'$ is a solution of $Ax = cb + c'b'$.

Example: The vectors $x = (1, 1, 2)$ and $x' = (0, 1, -1)$ are the respective solutions of the following systems.

$$\begin{aligned} x_1 - x_2 + x_3 &= 2 \\ 3x_1 + x_2 - x_3 &= 2 \\ 2x_1 + 2x_2 + x_3 &= 6 \\ x_1 - x_2 + x_3 &= -2 \\ 3x_1 + x_2 - x_3 &= 2 \\ 2x_1 + 2x_2 + x_3 &= 1 \end{aligned}$$

Note that the two systems have the same left side. The following systems

$$\begin{aligned} x_1 - x_2 + x_3 &= 0 \\ 3x_1 + x_2 - x_3 &= 4 \\ 2x_1 + 2x_2 + x_3 &= 7 \\ x_1 - x_2 + x_3 &= 1 \\ 3x_1 + x_2 - x_3 &= 1 \\ 2x_1 + 2x_2 + x_3 &= 3 \\ x_1 - x_2 + x_3 &= 8 \\ 3x_1 + x_2 - x_3 &= -4 \\ 2x_1 + 2x_2 + x_3 &= 3 \end{aligned}$$

still have the same left side but different right sides. By

$$\begin{aligned} (0, 4, 7) &= (2, 2, 6) + (-2, 2\ 1), \\ (1, 1, 3) &= 1/2\ (2, 2, 6), \\ (8, -4, 3) &= (2, 2, 6) - 3(-2, 2\ 1), \end{aligned}$$

we see that

$$\begin{aligned} x + x' &= (1, 1, 2) + (0, 1, -1) = (1, 2, 1), \\ 1/2\ x &= 1/2\ (1, 1, 2) = (1/2, 1/2, 1), \\ x - 3x' &= (1, 1, 2) - 3(0, 1, -1) = (1, -2, 5), \end{aligned}$$

are the respective solutions of the three new systems.

MATRIX TRANSFORMATIONS ARE LINEAR

Theorem $A(x + y) = Ax + Ay$, $A(cx) = cAx$.

Proof By an earlier exercise, we have

$x_1a_1 + x_2a_2 + ... + x_na_n = Ax$

where $A = [a_1\ a_2 ...\ a_n]$ is a matrix with the vectors $a_1, a_2, ..., a_n$ as columns, and $x_1, x_2, ..., x_n$ are the coordinates of x. Thus

$$\begin{aligned} Ax + Ay &= x_1a_1 + x_2a_2 + ... + x_na_n + y_1a_1 + y_2a_2 + ... + y_na_n \\ &= (x_1a_1 + y_1a_1) + (x_2a_2 + y_2a_2) + ... + (x_na_n + y_na_n) \\ &= (x_1 + y_1)a_1 + (x_2 + y_2)a_2 + ... + (x_n + y_n)a_n \end{aligned}$$

Since $x_1 + y_1, x_2 + y_2, \ldots, x_n + y_n$ are the coordinates of $x + y$, we see that the summation is $A(x + y)$.

The proof for $A(cx) = cAx$ is similar.

LINEARITY IN PICTURE

We can often visualize transformations between euclidean spaces. For example, the name of transformation *Reflection in x–axis*: $R^2 \to R^2$, $(x, y) \to (x, -y)$ is inspired by the picture.

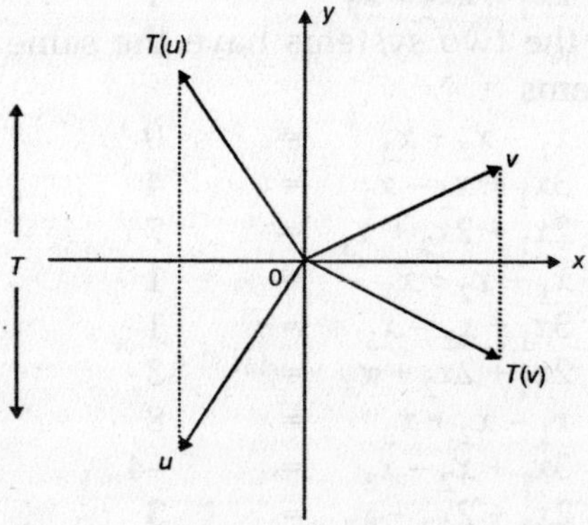

Example: The transformation *Reflection in y–axis*: $R^2 \to R^2$, $(x, y) \to (-x, y)$

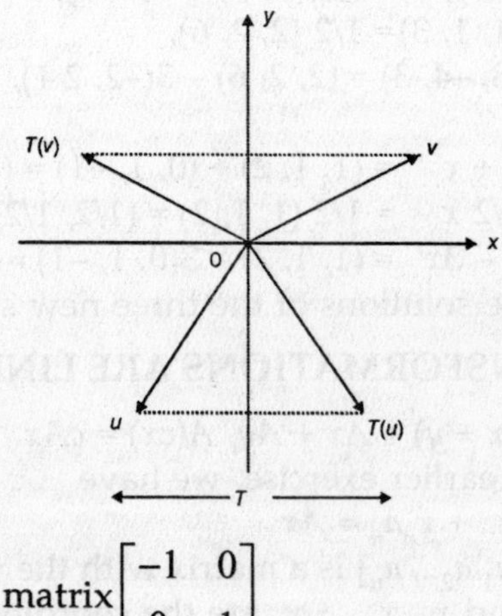

is given by the matrix $\begin{bmatrix} -1 & 0 \\ 0 & 1 \end{bmatrix}$

The transformation *Rotation by &\theta;*: $R^2 \to R^2$, $(x, y) \to (x\cos\theta + y\sin\theta, -x\sin\theta + y\cos\theta)$

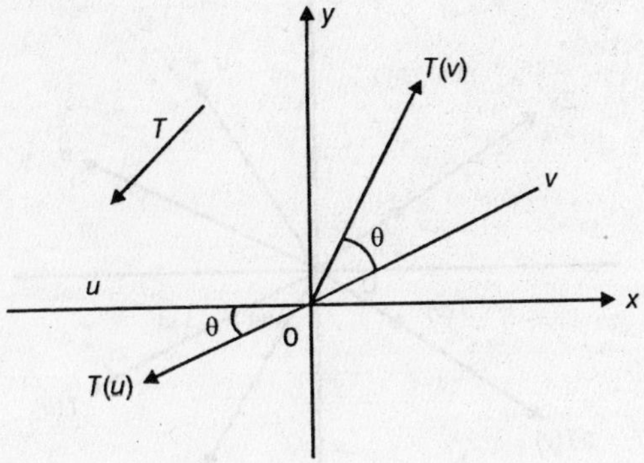

is given by the matrix (remember writing vectors vertically)

$$\begin{bmatrix} \cos\theta & -\sin\theta \\ \sin\theta & \cos\theta \end{bmatrix}$$

The transformation *Projection to* (x_1, x_3)*–plane*: $R^3 \rightarrow R^2$, $(x_1, x_2, x_3) \rightarrow (x_1, x_3)$

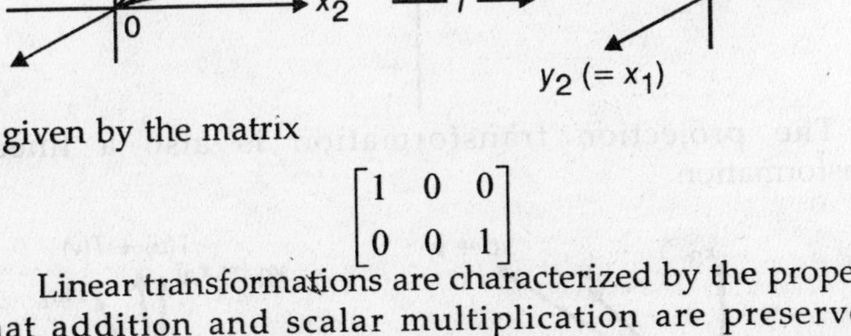

is given by the matrix

$$\begin{bmatrix} 1 & 0 & 0 \\ 0 & 0 & 1 \end{bmatrix}$$

Linear transformations are characterized by the property that addition and scalar multiplication are preserved. Geometrically, this means that parallelograms and stretching/shrinking are preserved.

Example: The transformation *Reflection in x–axis* clearly preserves parallelograms and stretching/shrinking.

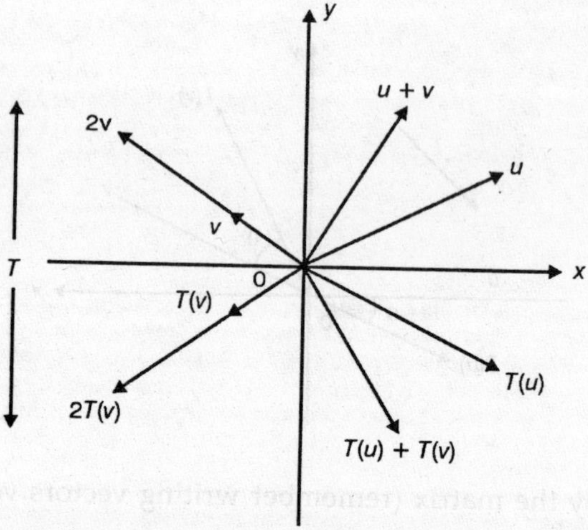

The rotation transformation also preserves the two operations.

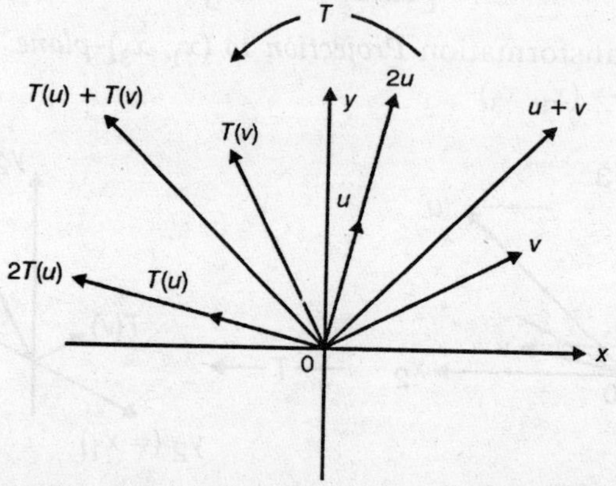

The projection transformation is also a linear transformation.

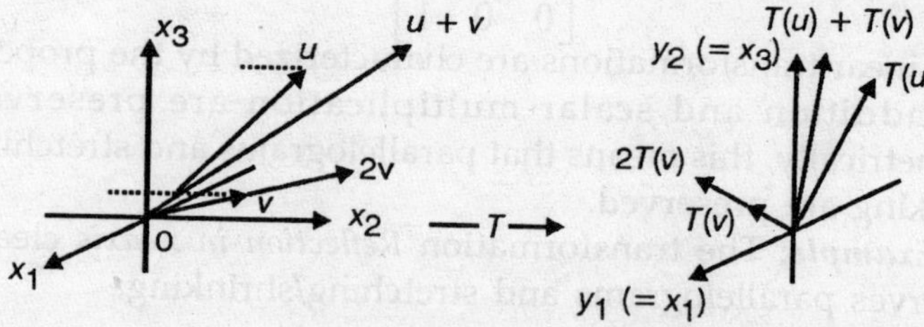

LINEAR TRANSFORMATION = MATRIX TRANSFORMATION

We have seen matrix transformations are linear. Conversely, this example suggests the converse is also true.

Any linear transformation T: $R^n \rightarrow R^m$ is also a matrix transformation, with the matrix given by

$$A = [T(e_1)\ T(e_2) \ldots\ T(e_n)].$$

In other words, the columns of the matrix A are the values of the linear transformation T at the standard basis. Thus the concepts of matrix transformations and linear transformations (between euclidean spaces) are equivalent.

Example: The identity transformation id: $R^n \rightarrow R^n$ is clearly linear. The corresponding matrix is

$$[id(e_1)\ id(e_2) \ldots\ id(e_n)] = [e_1\ e_2 \ldots\ e_n] = \begin{bmatrix} 1 & 0 & \ldots & 0 \\ 0 & 1 & \ldots & 0 \\ \vdots & \vdots & & \vdots \\ 0 & 0 & \ldots & 1 \end{bmatrix}$$

called the identity matrix and denoted by I_n.

Example: In this example, we started from a matrix

$$A = \begin{bmatrix} 1 & -1 & 1 \\ 3 & 1 & -1 \\ 2 & 2 & 1 \end{bmatrix}$$

and produced a matrix transformation (which is also a linear transformation).

$$T\begin{bmatrix} x_1 \\ x_2 \\ x_3 \end{bmatrix} = \begin{bmatrix} x_1 - x_2 + x_3 \\ 3x_1 + x_2 - x_3 \\ 2x_1 + 2x_2 + x_3 \end{bmatrix} \quad R^3 \rightarrow R^3$$

Conversely, we may recover the columns of A as follows

$$[\text{col } 1] \text{ of } A = T\begin{bmatrix}1\\0\\0\end{bmatrix} = \begin{bmatrix}1-0+0\\3+0-0\\0-1+0\end{bmatrix} = \begin{bmatrix}1\\3\\2\end{bmatrix}$$

$$[\text{col } 2] \text{ of } A = T\begin{bmatrix}0\\0\\1\end{bmatrix} = \begin{bmatrix}0-0-0\\0+1-0\\0-2+0\end{bmatrix} = \begin{bmatrix}-1\\1\\2\end{bmatrix}$$

$$[\text{col } 3] \text{ of } A = T\begin{bmatrix}0\\0\\1\end{bmatrix} = \begin{bmatrix}0-0+1\\0+0-1\\0-0+1\end{bmatrix}$$

Example: The reflections and rotations on R^2 are linear transformations because they preserve parallelogram and stretching/shrinking. To find their matrices, we simply need to apply reflections and rotations to the standard basis.

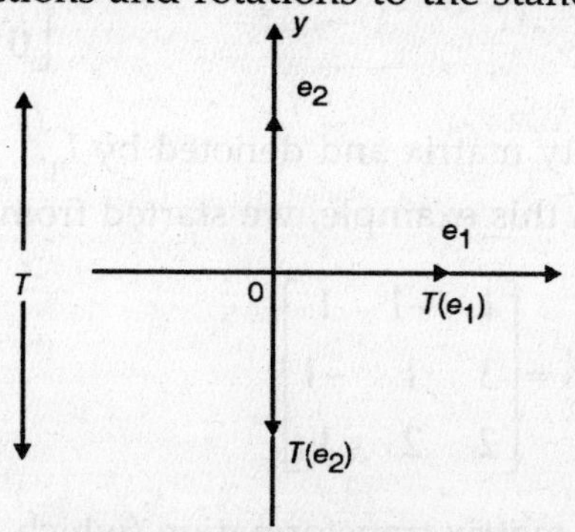

The picture shows that the reflections of e_1, e_2 in x–axis are $e_1 = (1, 0)$, $-e_2 = (0, -1)$. Therefore the matrix for the reflection is

$$\begin{bmatrix}1 & 0\\0 & -1\end{bmatrix}$$

Moreover, we also see from picture that for the rotation by angle θ, $T(e_1) = (\cos\theta, \sin\theta)$, $T(e_2) = (-\sin\theta, \cos\theta)$.

Therefore the matrix for the rotation is

$$\begin{bmatrix} \cos\theta & -\sin\theta \\ \sin\theta & \cos\theta \end{bmatrix}$$

Exercise: Find the corresponding transformations, and indicate the source and target euclidean spaces.

1. [1 2]

Ans. $T(x_1, x_2) = x_1 + 2x_2$: $R^2 \to R^1$

2. $\begin{bmatrix} 1 \\ 2 \end{bmatrix}$

Ans. $T(x_1) = (x_1, 2x_1)$: $R^1 \to R^2$

3. $\begin{bmatrix} 2 & 0 & 2 & 1 \\ 1 & 1 & -1 & 0 \\ 0 & 1 & -2 & 1 \end{bmatrix}$

Ans. $T(x_1, x_2, x_3, x_4) = (x_1 + 2x_3 + x_4, x_1 + x_2 - x_3, x2 - 2x_3 + x_4)$: $R^4 \to R^3$

4. $\begin{bmatrix} 1 & 2 & 1 \\ 2 & 1 & 1 \\ 0 & 3 & 1 \\ 3 & 0 & 1 \end{bmatrix}$

Ans. $T(x_1, x_2, x_3) = (x_1 + 2x_2 + x_3, 2x_1 + x_2 + x_3, 3x2 + x_3, 3x_1 + x_3)$: $R^3 \to R^4$

5. $\begin{bmatrix} 1 & -1 & 1 & 0 & 2 \\ 2 & -2 & 0 & 2 & 2 \\ -1 & 1 & 2 & -3 & 1 \\ -2 & 2 & 1 & -3 & -1 \end{bmatrix}$

Ans. $T(x_1, x_2, x_3, x_4, x_5) = (x_1 - x_2 + x_3 + 2x_5, 2x_1 - 2x_2 + 2x_4 + 2x_5, - x_1 + x_2 + 2x_3 - 3x_4 + x5, - 2x_1 + 2x_2 + 2x_3 - 3x_4 - x5)$: $R^5 \to R^4$

Exercise: Which of the following are matrix transformations? For the ones that are, find the corresponding matrices. For the ones that are not, provide the reason.

1. $T_1(x_1, x_2, x_3) = (2x_1 + 3x_2 - 7x_3, 10x_1 - 4x_2 - 8x_3, - 4x_1 + x_2)$

Ans. This is a matrix transformation.

$$\begin{bmatrix} 2 & 3 & -7 \\ 10 & -4 & -8 \\ -4 & 1 & 0 \end{bmatrix}$$

2. $T_2(x_1, x_2, x_3) = (2x_1 + 3x_2 - 7x_3 + 1, 10x_1 - 4x_2 - 8x_3, - 4x_1 + x_2 - 5)$

Ans. This is not a matrix/linear transformation because $T_2(0, 0, 0) = (1, 0, -5)$, whereas any linear transformation must send zero vector to zero vector.

3. $T_3(x_1, x_2, x_3) = 2x_1 + 3x_2 - 7x_3$

Ans. This is a matrix transformation.

$$\begin{bmatrix} 2 & 3 & -7 \end{bmatrix}$$

4. $T_4(x_1, x_2, x_3) = 2x_1 + 3x_2 - 7x_3 + 1$

Ans. This is not a matrix/linear transformation because $T_4(0, 0, 0) = 1$, which is nonzero.

5. $T_5(x_1, x_2, x_3) = (2x_1, 3x_2, - 7x_3)$

Ans. This is a matrix transformation.

$$\begin{bmatrix} 2 & 0 & 0 \\ 0 & 3 & 0 \\ 0 & 0 & -7 \end{bmatrix}$$

6. $T_6(x_1, x_2, x_3) = (2x_2, 3x_3, - 7x_1)$

Ans. This is a matrix transformation.

$$\begin{bmatrix} 2 & 0 & 0 \\ 0 & 0 & 3 \\ -7 & 0 & 0 \end{bmatrix}$$

7. $T_7(x_1, x_2, x_3) = (3x_2, - 7x_3, 2x_1)$

Ans. This is a matrix transformation.

$$\begin{bmatrix} 0 & 3 & 0 \\ 0 & 0 & -7 \\ 2 & 0 & 0 \end{bmatrix}$$

8. $T_8(x) = (2x, 3x, -7x)$

Ans. This is a matrix transformation.

$$\begin{bmatrix} 2 \\ 3 \\ -7 \end{bmatrix}$$

9. $T_9(x, y, z) = (x - y, y - z, z - x)$

Ans. This is a matrix transformation.

$$\begin{bmatrix} 1 & -1 & 0 \\ 0 & 1 & -1 \\ -1 & 0 & 1 \end{bmatrix}$$

10. $\boldsymbol{T}_{10}(x, y) = (x + y, xy)$

Ans. This is not a matrix/linear transformation because $T_{10}(1, 0) = T_{10}(0, 1) = (1,0)$, $T_{10}(1, 1) = (2, 1) \rightarrow T_{10}(1, 0) + T_{10}(0, 1)$.

11. $T_{11}(x, y) = (x + y, 2^{1/2}x + 3^{1/2}y)$

Ans. This is a matrix transformation.

$$\begin{bmatrix} 1 & 1 \\ 2^{1/2} & 3^{1/2} \end{bmatrix}$$

12. $\boldsymbol{T}_{12}(u, v) = (u + v, u^2 + v^2, u^3 + v^3)$

Ans. This is not a matrix/linear transformation because $T_{12}(1, 1) = (2, 2, 2)$, $T_{12}(2, 2) = (4, 8, 16)$, so that $T_{12}(2, 2)$ '" $2T_{12}(1, 1)$.

Exercise: Convince yourself that the following transformations are linear by drawing the picture and indicating how the parallelogram and scaling by 1/2 are preserved by the transformations. Then find the corresponding matrices.

1. *Reflection in diagonal line*: $R^2 \to R^2$

Ans. The following is the picture for the transformation.

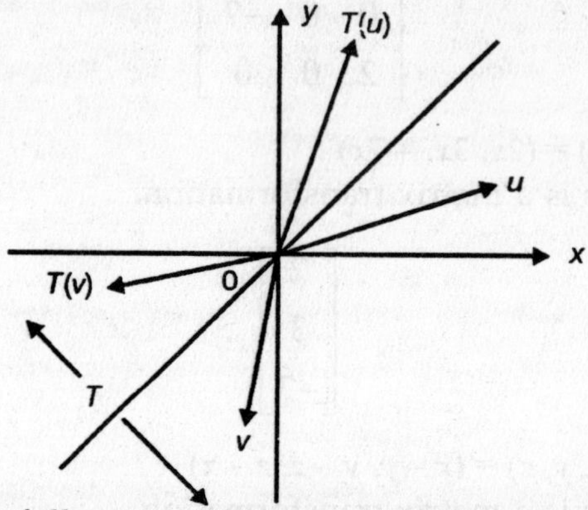

The following illustrates that the transformation is linear.

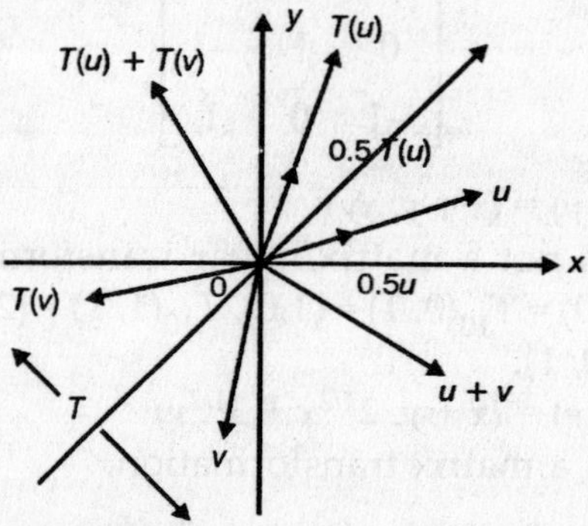

From this picture,

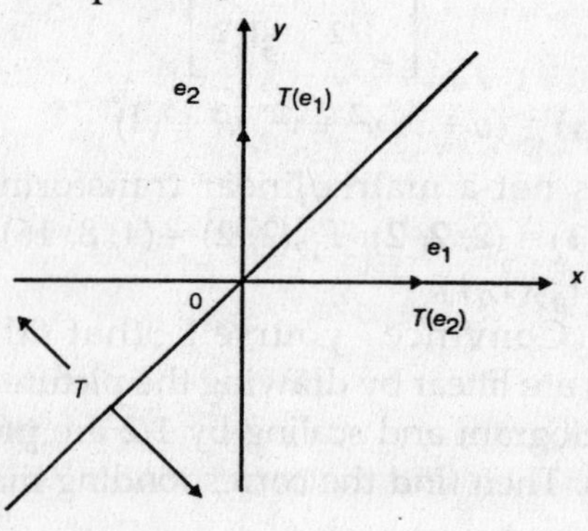

we see $T(e_1) = e_2 = (0, 1)$, $T(e_2) = e_1 = (1, 0)$. Therefore the matrix is

$$\begin{bmatrix} 0 & 1 \\ 1 & 0 \end{bmatrix}$$

2. *Projection to x_1–line*: $R^2 \to R^1$

Ans. The following is the picture for the transformation.

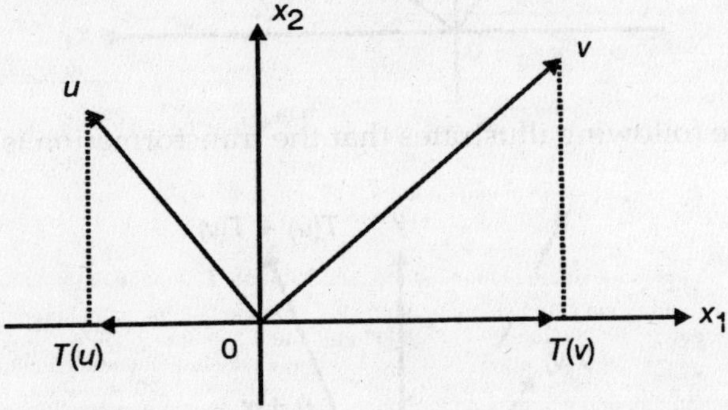

The following illustrates that the transformation is linear. We clearly have $T(e_1) = e_1 = (1, 0)$, $T(e_2) = (0, 0)$. Therefore the matrix is

$$\begin{bmatrix} 1 & 0 \\ 0 & 0 \end{bmatrix}$$

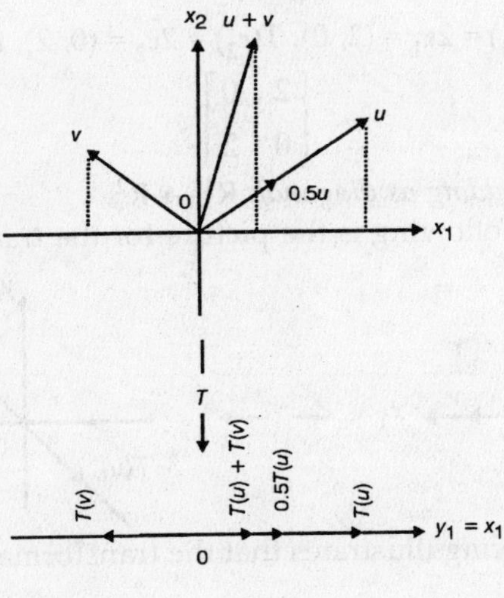

3. *Multiplying by* 2: $R^2 \to R^2$

Ans. The following is the picture for the transformation.

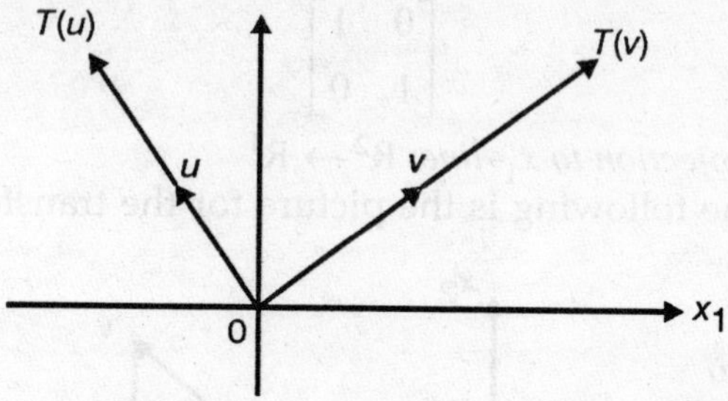

The following illustrates that the transformation is linear.

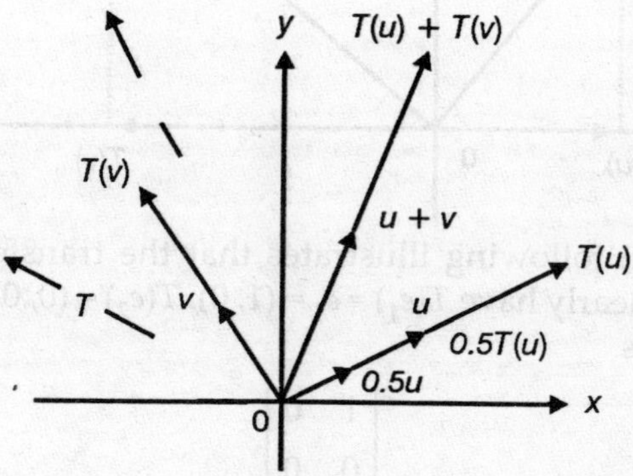

Since $T(e_1) = 2e_1 = (2, 0)$, $T(e_2) = 2e_2 = (0, 2)$, the matrix is

$$\begin{bmatrix} 2 & 0 \\ 0 & 2 \end{bmatrix}$$

4. *Embedding as diagonal*: $R^1 \to R^2$

Ans. The following is the picture for the transformation.

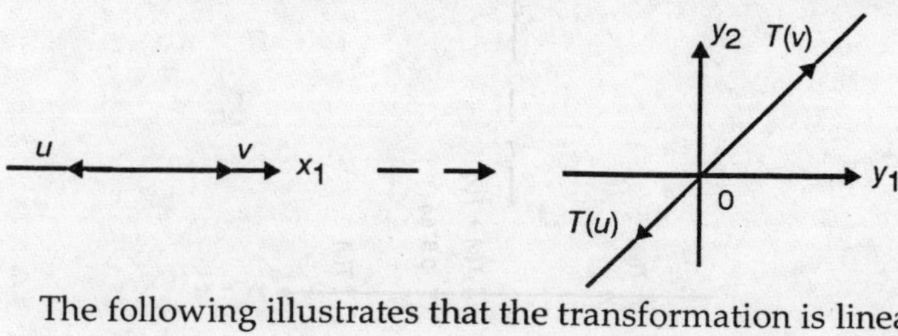

The following illustrates that the transformation is linear.

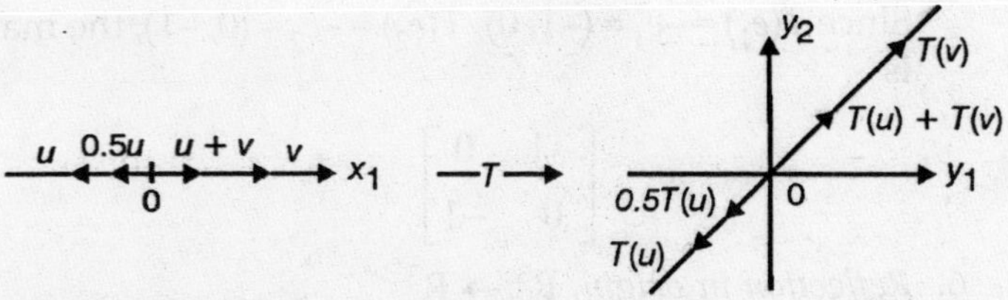

Since $T(e_1) = (1, 1)$ (note that $e_1 = 1$ is a 1–dimensional vector), the matrix is

$$\begin{bmatrix} 1 \\ 1 \end{bmatrix}$$

5. *Reflection in origin*: $R^2 \to R^2$

Ans. The following is the picture for the transformation.

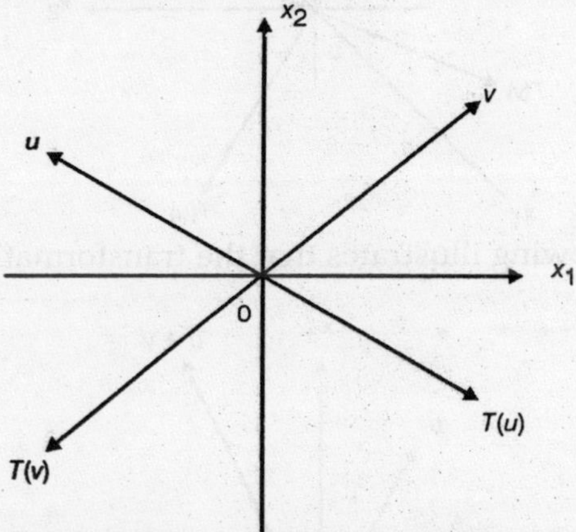

The following illustrates that the transformation is linear.

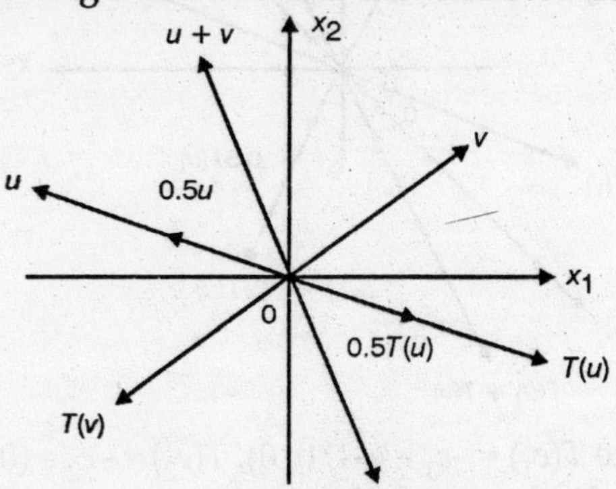

Since $T(e_1) = -e_1 = (-1, 0)$, $T(e_2) = -e_2 = (0, -1)$, the matrix is

$$\begin{bmatrix} -1 & 0 \\ 0 & -1 \end{bmatrix}$$

6. *Reflection in origin*: $R^3 \to R^3$

Ans. The following is the picture for the transformation.

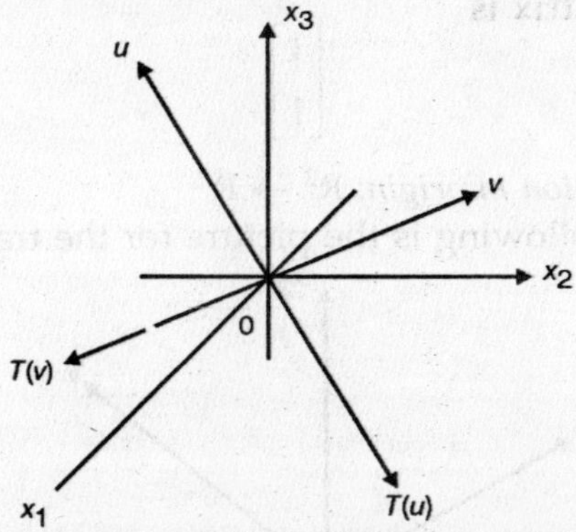

The following illustrates that the transformation is linear.

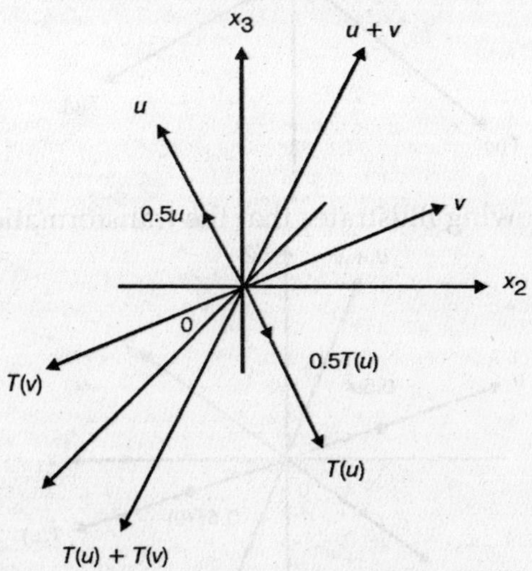

Since $T(e_1) = -e_1 = (-1, 0, 0)$, $T(e_2) = -e_2 = (0, -1, 0)$, $T(e_3) = -e_3 = (0, 0, -1)$, the matrix is

$$\begin{bmatrix} -1 & 0 & 0 \\ 0 & -1 & 0 \\ 0 & 0 & -1 \end{bmatrix}$$

Exercise: The values of the following linear transformations at the standard basis have been given. Find the value at the other vectors.

Ans.

1. $T: R^2 \to R^2$, $T(e_1) = (5, -3)$, $T(e_2) = (-4, 2)$, find $T(2, 3)$, $T(-3, 1)$, $T(x_1, x_2)$.

Ans. By

$$(2, 3) = 2e_1 + 3e_2$$
$$(-3, 1) = -3e_1 + e_2$$
$$(x_1, x_2) = x_1e_1 + x_2e_2$$

We have

$T(2, 3) = 2T(e_1) + 3T(e_2) = 2(5, -3) + 3(-4, 2) = (-2, 0)$ $T(-3, 1) = -3T(e_1) + T(e_2) = -3(5, -3) + (-4, 2) = (-19, 9)$

$T(x_1, x_2) = x_1T(e_1) + x_2T(e_2) = x_1(5, -3) + x_2(-4, 2)$
$= (5x_1 - 4x_2, -3x_1 + 2x_2)$

2. $S: R^2 \to R^3$, $S(e_1) = (1, 2, 3)$, $S(e_2) = (4, 5, 6)$, find $S(2, 3)$, $S(-3, 1)$, $S(x_1, x_2)$.

Ans. By

$$(2, 3) = 2e_1 + 3e_2$$
$$(-3, 1) = -3e_1 + e_2$$
$$(x_1, x_2) = x_1e_1 + x_2e_2$$

We have

$S(2, 3) = 2S(e_1) + 3S(e_2) = 2(1, 2, 3) + 3(4, 5, 6) = (14, 19, 24)$

$S(-3, 1) = -3S(e_1) + S(e_2) = -3(1, 2, 3) + (4, 5, 6) = (1, -1, -3)$

$S(x_1, x_2) = x_1S(e_1) + x_2S(e_2) = x_1(1, 2, 3) + x_2(4, 5, 6)$
$= (x_1 + 4x_2, 2x_1 + 5x_2, 3x_1 + 6x_2)$

3. $R: R^3 \to R^2$, $R(e_1) = (1, 2)$, $R(e_2) = (3, 4)$, $R(e3) = (5, 6)$, find $R(1, 0, 2)$, $R(-2, 1, 3)$, $R(x_1, x_2, x_3)$.

Ans. By

$$(1, 0, 2) = e_1 + 2e_3$$
$$(-2, 1, 3) = -2e_1 + e_2 + 3e_3$$

$(x_1, x_2, x_3) = x_1e_1 + x_2e_2 + x_3e_3$

We have

$R(1, 0, 2) = R(e_1) + 2S(e_3) = (1, 2) + 2(5, 6) = (11, 14)$

$R(-2, 1, 3) = -2R(e_1) + R(e_2) + 3R(e_3)$

$= -2(1, 2) + (3, 4) + 3(5, 6) = (16, 18)$

$R(x_1, x_2, x_3) = x_1R(e_1) + x_2R(e_2) + x_3R(e3) = x_1(1, 2) + x_2(3, 4) + x_3(5, 6)$

$= (x_1 + 3x_2 + 5x_3, 2x_1 + 4x_2 + 6x_3)$

Exercise: Let T: $R^2 \rightarrow R^3$ be a linear transformation. Assume we already know $T(1, 2) = (1, 2, 3)$, $T(3, 4) = (4, 5, 6)$.

- Use this exercise and the linearity of T to compute $T(4, 6)$, $T(1, 1)$, $T(1, 0)$, $T(0, 1)$.
- Use $T(1, 0)$, $T(0, 1)$ to compute $T(x_1, x_2)$ in general.
- Use $3(1, 1, 1) = (4, 5, 6) - (1, 2, 3)$ to find x_1, x_2 such that $T(x_1, x_2) = (1, 1, 1)$

Answer: From the exercise, we have

$(4, 6) = (1, 2) + (3, 4)$

$(1, 1) = -1/2\ (1, 2) + 1/2\ (3, 4)$

$(1, 0) = -2(1, 2) + (3, 4)$

$(0, 1) = 3/2\ (1,2) - 1/2\ (3, 4)$

Thus

$T(4, 6) = T(1, 2) + T(3, 4) = (1, 2, 3) + (4, 5, 6) = (5, 7, 9)$

$T(1, 1) = -1/2\ T(1, 2) + 1/2\ T(3, 4) = -1/2\ (1, 2, 3) + 1/2\ (4, 5, 6) = (3/2, 3/2. 3/2)$

$T(1, 0) = -2T(1, 2) + T(3, 4) = -2(1, 2, 3) + (4, 5, 6) = (2, 1, 0)$

$T(0, 1) = 3/2\ T(1,2) - 1/2\ T(3, 4) = 3/2\ (1, 2, 3) - 1/2\ (4, 5, 6) = (-1/2, 1/2, 3/2)$

Moreover,

$T(x_1, x_2) = x_1T(1, 0) + x_2T(0, 1) = x_1(2, 1, 0) + x_2(-1/2, 1/2, 3/2) = (2x_1 - 1/2x_2, x_1 + 1/2x_2, 3/2x_2)$.

Finally, from the equality $3(1, 1, 1) = (4, 5, 6) - (1, 2, 3) = T(3, 4) - T(1, 2) = T(2, 2)$ we have $(1, 1, 1) = T(2/3, 2/3)$. Therefore for the last question, we may take $x_1 = x_2 = 2/3$.

Exercise: Let T: R^3 '! R^2 be a linear transformation. Assume we already know $T(1, 1, 1) = (1, 2)$, $T(1, 1, 0) = (3, 4)$, $T(1, 0, 0) = (5, 6)$. Find the general formula (or the matrix) for T.

Ans: We need to find $T(x_1, x_2, x_3)$. We use the same idea as in the previous exercise. First we solve the three systems $c_1(1, 1, 1) + c_2(1, 1, 0) + c3(1, 0, 0) = e_i$ for $i = 1, 2, 3$ and find

$$e_1 = (1, 0, 0)$$
$$e_2 = (1, 1, 0) - (1, 0, 0)$$
$$e_3 = (1, 1, 1) - (1, 1, 0)$$

Then we use the linearity of T to get

$$T(e_1) = T(1, 0, 0) = (5, 6)$$
$$T(e_2) = T(1, 1, 0) - T(1, 0, 0) = (3, 4) - (5, 6) = (-2, -2)$$
$$T(e_3) = T(1, 1, 1) - T(1, 1, 0) = (1, 2) - (3, 4) = (-2, -2)$$

Finally we use $(x_1, x_2, x_3) = x_1e_1 + x_2e_2 + x_3e_3$ and the linearity of T to get

$T(x_1, x_2, x_3) = x_1T(e_1) + x_2T(e_2) + x_3T(e_3) = x_1(5, 6) + x_2(-2, -2) + x_3(-2, -2) = (5x_1 - 2x_2 - 2x_3, 6x_1 - 2x_2 - 2x_3)$.

Index

T

V